Friction Stir Superplasticity for Unitized Structures

Friction Stir Superplasticity for Unitized Structures

A volume in the *Friction Stir Welding and Processing Book Series*

Zongyi Ma
Institute of Metal Research, Chinese Academy of Sciences, Shenyang, China

Rajiv S. Mishra
Department of Materials Science and Engineering, University of North Texas, Denton, TX, USA

ELSEVIER

AMSTERDAM • BOSTON • HEIDELBERG • LONDON
NEW YORK • OXFORD • PARIS • SAN DIEGO
SAN FRANCISCO • SINGAPORE • SYDNEY • TOKYO
Butterworth-Heinemann is an imprint of Elsevier

BH

Butterworth-Heinemann is an imprint of Elsevier
225 Wyman Street, Waltham, MA 02451, USA
The Boulevard, Langford Lane, Kidlington, Oxford, OX5 1GB, UK

Library of Congress Cataloging-in-Publication Data
A catalog record for this book is available from the Library of Congress

British Library Cataloguing in Publication Data
A catalogue record for this book is available from the British Library

ISBN: 978-0-12-420006-7

For information on all Butterworth-Heinemann publications visit our website at **store.elsevier.com**

This book has been manufactured using Print On Demand technology. Each copy is produced to order and is limited to black ink. The online version of this book will show color figures where appropriate.

CONTENTS

PREFACE

This is the third volume in the recently launched short book series on friction stir welding and processing. As highlighted in the preface of the first book, the intention of this book series is to serve engineers and researchers engaged in advanced and innovative manufacturing techniques. Friction stir processing was started as a generic microstructural modification technique almost 15 years back. In that period, friction stir processing related research has shown wide promise as a versatile microstructural modification technique and solid state manufacturing technology. Yet, broader implementations have been sorely missing. Disruptive technologies face greater barrier to implementation as designers do not have these in their traditional design tool box! Part of the inhibition is due to lack of maturity and availability of large data set.

This book is focused on grain refinement induced enhancement in superplastic properties, which was the first demonstration of friction stir processing in 1999. The nature of this short book for the most part is that of a monograph. The authors have used large volume of information from their published work. It includes discussion on possibilities of expanding the domain of superplastic forming by using friction stir processing as an enabling technology. Currently designers do not consider superplastic forming of thick parts as that was not possible before. Hopefully this book and increasing volume of published literature will provide confidence to designers and engineers to consider unitized structures outside the current design limitations. As stated in the previous volume, this short book series on friction stir welding and processing will include books that advance both the science and technology.

Rajiv S. Mishra
University of North Texas
April 29, 2014

ACKNOWLEDGMENTS

The authors would like to sincerely thank all that have provided contributions that have made this short book possible. Much of the work was done over the period of 1999–2008 and a number of students and postdoctoral researchers helped in the initial stages of this effort (Indrajit Charit, Zongyi Ma, Siddharth Sharma, Abhijit Dutta, Lucie Johannes, and Yanwen Wang). This research area got started with encouragement and mentorship from Murray Mahoney and Amiya Mukherjee.

Ma gratefully acknowledges the support of the National Natural Science Foundation of China under Grant Nos. 50671103, 50871111, and 51331008, the National Outstanding Young Scientist Foundation with Grant No. 50525103 and the Hundred Talents Program of Chinese Academy of Sciences.

Mishra gratefully acknowledges the broader support of the National Science Foundation for the NSF-IUCRC for Friction Stir Processing (NSF-IIP-0531019 and NSF-IIP-1157754) and the support of its industrial members. It has allowed him to sustain collaborative work and writing. The support from the National Science Foundation through NSF-CMMI-0085044 and NSF-CMMI-0323725 was critical in the development of initial results presented in this book.

CHAPTER 1

Introduction

Superplasticity refers to the ability of materials to exhibit high uniform elongation when pulled in tension while maintaining a stable microstructure. Superplastic forming (SPF) of commercial aluminum alloys has been considered as one of the important fabrication methods for unitized components in automotive and aerospace industries. There is increasing interest in these industries to implement wider use of SPF of commercial aluminum alloys in fabricating complex parts.

A prerequisite for achieving structural superplasticity is a fine-grain size, typically less than 15 μm. Conventionally, elaborate thermomechanical processing (TMP) is needed to obtain a fine microstructure conductive to superplastic deformation. For example, Paton et al. [1] developed a four-step TMP treatment to obtain grain size in the range of 8–14 μm in commercial 7075 and 7475 aluminum alloys. Jiang et al. [2,3] suggested an improved TMP approach for 7075 Al alloy that involved solution treatment, overaging, multiple pass warm rolling (200–220°C) with intermittent reheating and a recrystallization treatment. Clearly, TMP for fine-grain microstructures is complex and time-consuming and leads to increased material cost. Furthermore, the optimum superplastic strain rate of $1-10 \times 10^{-4}\ s^{-1}$ obtained in these aluminum alloys [1–3] is too slow for superplastic forging/forming of components in the automotive industry.

From the viewpoint of practical industrial fabrication, it is highly desirable to perform SPF at higher strain rate and/or lower temperature. A higher forming rate of greater than $1 \times 10^{-2}\ s^{-1}$ would satisfy the current industrial fabrication speed [4]. On the other hand, a lower forming temperature would save energy, prevent grain growth, and reduce cavitation level and solute loss from surface layer, thereby maintaining superior post-forming properties [5]. To advance SPF into production-oriented industries, there is a need to develop new processing techniques to shift the optimum superplastic strain rate to high strain rate ($>10^{-2}\ s^{-1}$).

Constitutive relationship for superplasticity of fine-grained aluminum alloys predicts that a decrease in the grain size results in an increase in optimum superplasticity strain rate and a decrease in optimum superplasticity temperature, which have been verified by a number of experimental investigations [6–8]. In the past few years, numerous research efforts have been focused on development of fine-grained aluminum alloys exhibiting high-strain-rate superplasticity (HSRS) or low-temperature superplasticity (LTSP) by using thermo-mechanical treatment (TMT) [9–11], equal channel angular pressing (ECAP) [12–16], high-pressure torsion (HPT) [17], multiaxial alternative forging (MAF) [18], and accumulative roll bonding (ARB) [19]. However, these processing routes are subject to either higher material cost due to complex procedures and/or equipments or limitation of processed samples' size and shape. Clearly, there is a need for economical processing techniques that can produce the microstructure amenable to HSRS in commercial alloys.

Friction stir processing (FSP) is a novel plastic working technique [20], developed based on friction stir welding (FSW), an innovative solid joining technique invented in The Welding Institute in UK [21]. FSP generates significant frictional heating and intense plastic deformation, thereby resulting in occurrence of recrystallization in the stirred zone. In this case, a fine-grain region was produced in the FSP samples. Depending on tool geometry, FSP parameters, material chemistry, workpiece temperature, varied fine-grained microstructures were produced by FSP.

Mishra et al. [22,23] reported the first result for HSRS of fine-grained 7075 Al alloy prepared via FSP. Subsequently, superplasticity was obtained in a series aluminum alloys, such as 7075 Al, 2024 Al, 5083 Al, A356, Al–Mg–Zr, and Al–Zn–Mg–Sc [24–31]. These results indicate that FSP is a very effective technique to induce superplasticity in metal materials such as aluminum alloys and magnesium alloys. Table 3.1 summarizes the superplastic elongation observed in a number of aluminum alloys prepared via FSP. In this book, the current state of understanding and development of the superplastic behavior of FSP aluminum alloys is presented. Particular emphasis has been given to: (i) effects of FSP parameters on resultant microstructure and final superplastic properties and (ii) mechanisms responsible for the superplastic deformation, in particular at high strain rate and low temperature.

CHAPTER 2

Friction Stir Microstructure for Superplasticity

The microstructural prerequisites for a superplastic material are well established in metallic alloys. The first prerequisite is a fine-grain size, typically less than 15 μm. The optimum strain rate for superplasticity increases with decreasing grain size when grain boundary sliding (GBS) is the dominant process. For a given strain rate, the finer grain sizes lead to lower flow stresses, which is beneficial for practical forming operation. The second prerequisite is thermal stability of the fine-grained microstructure at high temperatures. Single-phase materials generally do not show superplasticity because grain growth occurs rapidly at elevated temperatures. Hence, presence of second phases at grain boundaries is required to resist excessive grain growth. An appropriate amount of fine, uniformly distributed, and thermally stable second-phase particles is necessary to keep stable microstructure during superplastic deformation. The third prerequisite is high grain boundary misorientation. High-angle grain boundaries (HAGBs, angle ≥15°), particularly random ones promote GBS, whereas low-angle grain boundaries are generally believed to be not suitable for GBS. The fourth prerequisite is equiaxed grain shape. With equiaxed grains, grain boundaries can experience shear stress easily promoting GBS. The fifth prerequisite is mobility of grain boundaries. During GBS, stress concentration could be produced at various grain boundary discontinuities such as triple points. The migration of grain boundaries could lead to reduction in stress concentration. Thus, GBS can continue as a major deformation mechanism.

Fine-grained microstructure in friction stir processed (FSP) aluminum alloys is produced through dynamic recrystallization [19]. Therefore, the grains in FSP aluminum alloys generally exhibit equiaxed shape with quite uniform size, and they can be adjusted by changing tool geometry, FSP parameters, and cooling condition. Figure 2.1 shows typical optical micrographs of FSP 7075Al samples processed under different FSP parameters. It is clear that the FSP 7075Al samples exhibit fine and equiaxed grains and the size of the grains decrease from 7.5 to 3.8 μm as the thermal input of FSP decreases by reducing the rotation-rate/travel-speed ratio. It is well documented that in aluminum alloys processed via FSP, fine-grain size of 0.7–10 μm can be easily obtained. The grain size in the FSP aluminum alloys is within the grain size range required for attaining structural superplasticity. Therefore, it is expected that the FSP fine-grained aluminum alloys would exhibit good superplastic properties.

Furthermore, fine-grained microstructure in FSP aluminum alloys is characterized by a high fraction of HAGBs. Figure 2.2 shows grain boundary misorientation distribution in FSP 7075Al and 2024Al alloys. Clearly, the fraction of HAGBs is as high as 85–95% [23,29]. This ratio is significantly higher than that obtained in fine-grained aluminum alloys, processed by other plastic working methods, with a typical ratio of 50–65% [9–18]. Much higher ratio of HAGBs in the FSP aluminum alloys than in the fine-grained aluminum alloys prepared by other processing techniques is mainly attributed to the occurrence of complete

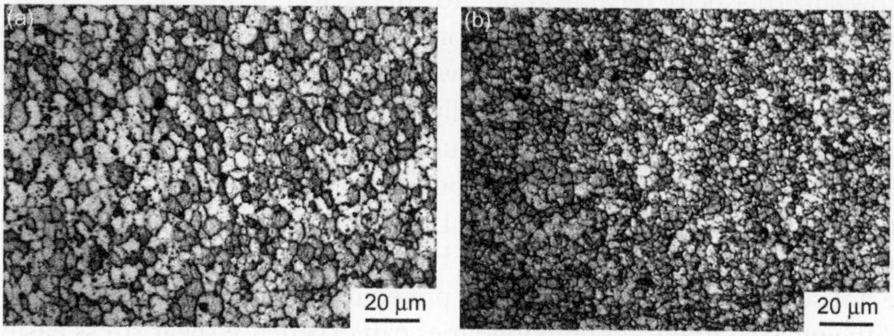

Figure 2.1 Optical micrographs showing grain structure of FSP 7075Al samples prepared under tool rotation-rate/travel-speed of (a) 400 rpm-102 mm/min (average grain size 7.5 μm) and (b) 350 rpm-154 mm/min (average grain size 3.8 μm) [24].

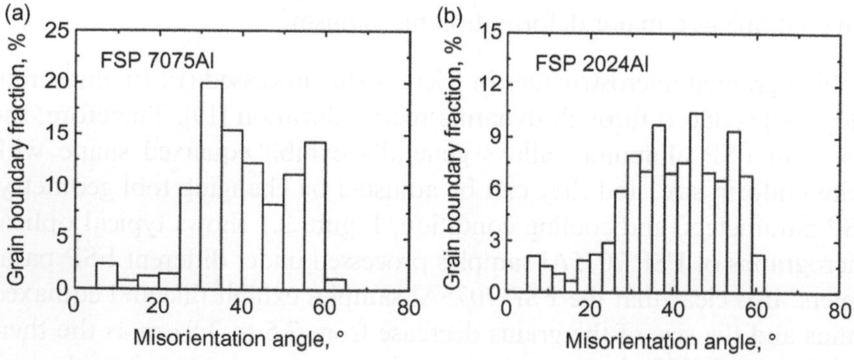

Figure 2.2 Grain boundary misorientation distribution in (a) FSP 7075Al and (b) FSP 2024Al, indicating the formation of a larger fraction of high-angle grain boundaries [23,29].

dynamic recrystallization during FSP. Predominant HAGBs is believed to be beneficial to superplasticity because HAGBs promote GBS.

Figure 2.3 shows the transmission electron microscopic (TEM) micrographs of FSP 7075Al-T651 samples with different grain sizes. Three important observations can be made. Firstly, the fine particles were uniformly distributed within the interior of grains and at the grain boundaries for both samples. Secondly, while the grain boundary particles usually exhibited a needle or disk-type morphology, the particles inside the grains generally had an equiaxed shape. Thirdly, the particles in the two FSP 7075Al alloys were fine and generally had a size of less than 0.5 μm. However, the FSP 3.8 μm-7075Al alloy exhibited smaller size and higher density of particles than the FSP 7.5 μm-7075Al alloy. These particles were identified as Cr-bearing dispersoids and $MnZn_2$-type precipitates [32]. The Cr-bearing dispersoids are expected to exert significant pinning effect on the grain boundaries.

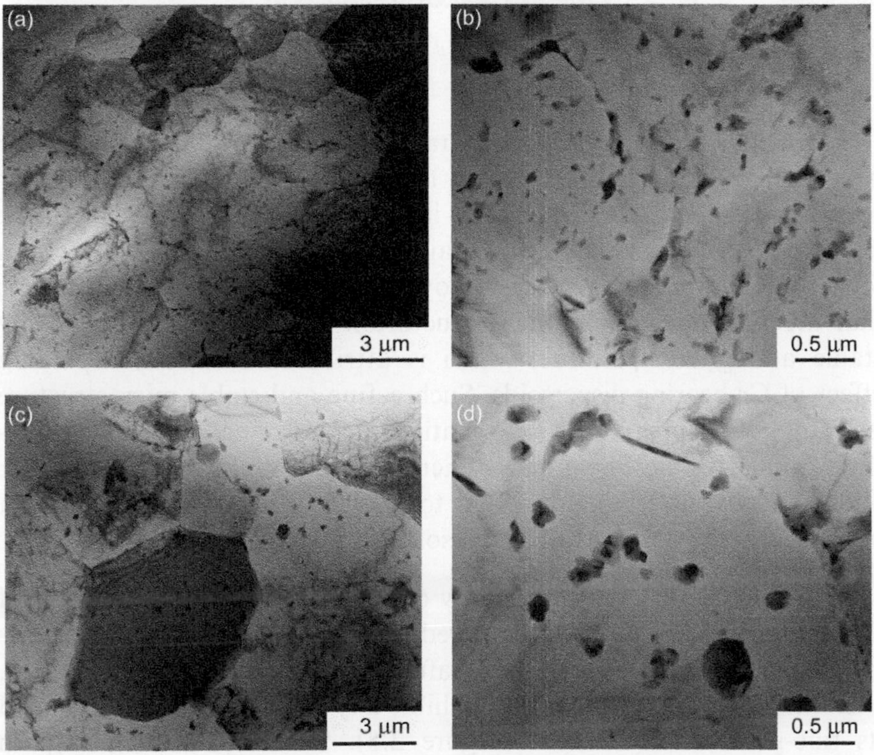

Figure 2.3 TEM micrographs showing distribution of precipitates in (a) and (b) FSP 3.8 μm-7075Al alloy, and (c) and (d) FSP 7.5 μm-7075Al alloy [140].

Figure 2.4 Optical micrographs showing grain structure of FSP 7075Al-T651: (a) 7.5 μm-7075Al, heat-treated at 490° C/1 h and (b) 3.8 μm-7075Al, heat-treated at 490° C/1 h [24].

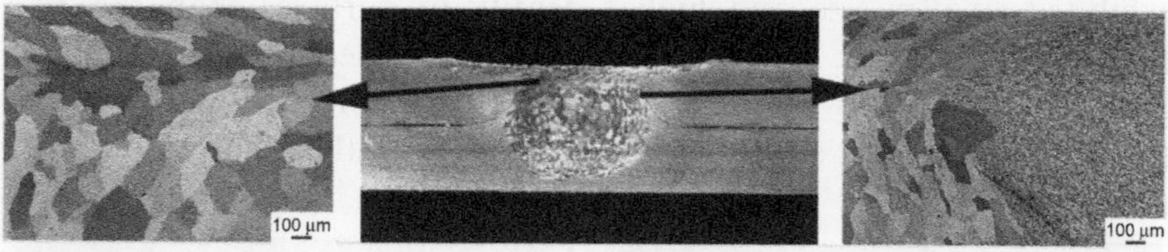

Figure 2.5 Optical micrographs for different regions of the single-pass FSW 7475Al lap joint annealed at 743 K for 30 min [143].

Figure 2.4 shows optical micrographs of FSP 7075Al samples under heat-treated conditions. Compared to the as-FSP microstructure, heat treatment at 490°C for 1 h resulted in slight coarsening of grains in the heat-treated samples. The mean grain size increased to 9.1 and 5.9 μm for FSP 7.5 and 3.5 μm-7075Al alloys, respectively. This demonstrates that the FSP fine-grained microstructure in 7075Al alloys is relatively stable at high temperature, which is attributed to effective pinning effect of Cr-bearing dispersoids. Such a fine and stable microstructure is suitable for superplastic deformation and forming at high temperature. Similarly, the fine-grained microstructure of FSP Al–4Mg–1Zr alloy is stable at temperatures up to 550°C due to the presence of a higher amount of fine Al_3Zr dispersoids of 20 nm.

However, a heat-treatment study on FSW 7475Al lap joint revealed that abnormal grain growth occurred within the stirred zone with fine-grained microstructure of 2.2 μm after annealing at 470°C for 0.5 h, whereas parent material did not exhibit any grain growth and retained its fine-grain microstructure (Figure 2.5). Similarly, abnormal grain growth was also observed at high temperature in FSP 7050 and 2519 aluminum alloys [33].

CHAPTER 3

High-Strain-Rate Superplasticity

High-strain-rate superplasticity (HSRS) refers to the superplasticity achieved at an optimum strain rate of $\geq 1 \times 10^{-2}\ s^{-1}$ [34]. According to the prediction by the constitutive relationship for superplasticity, the optimum strain rate increased with decreasing the grain size. Based on the fine-grain size of 0.7–10 μm produced in aluminum alloys via friction stir processing (FSP), it is expected that the FSP fine-grained aluminum alloys would exhibit high-strain-rate superplastic behavior with the decrease in the grain size. Table 3.1 summarizes the superplastic data of a number of aluminum alloys prepared by FSP. It is indicated that HSRS has been achieved in several FSP aluminum alloys, such as 7075Al, 2024Al, Al–Mg–Zr, Al–Zn–Mg–Sc, and Al–Mg–Sc.

3.1 SUPERPLASTIC BEHAVIOR

Figure 3.1(a) shows the variation of elongation with initial strain rate for as-rolled and as-FSP 7075Al alloys. As-rolled 7075Al alloy did not exhibit superplastic elongation at 480–490°C for initial strain rate of 1×10^{-3}–$1 \times 10^{-1}\ s^{-1}$. FSP resulted in generation of significant superplasticity in 7075Al alloy. For the FSP 7.5 μm-7075Al, an optimum strain rate of $3 \times 10^{-3}\ s^{-1}$ for maximum elongation was observed. A decrease in the grain size from 7.5 to 3.8 μm resulted in significantly enhanced superplasticity and a shift to higher optimum strain rate. Superplastic elongations >1250% were obtained in the strain rate range of 3×10^{-3}–$3 \times 10^{-2}\ s^{-1}$. Even at a high initial strain rate of $1 \times 10^{-1}\ s^{-1}$, an elongation of 750% was attained demonstrating HSRS.

Figure 3.1(b) shows the effect of temperature on the superplastic ductility of the FSP 7075Al at initial strain rates of 3×10^{-3} and $1 \times 10^{-2}\ s^{-1}$. Compared to the FSP 7.5 μm-7075Al, the FSP 3.8 μm-7075Al exhibited significantly enhanced superplasticity over a wide temperature range of 420–510°C, a shift to higher optimum superplastic strain rates, and a lower optimum superplastic deformation temperature. Further, even at the high temperature of 530°C, the FSP 7.5 μm and

Table 3.1 Summary of Superplastic Elongation Observed in a Number of Aluminum Alloys

Alloy	Temperature, °C	Strain rate, s^{-1}	Elongation, %
7075	480	1×10^{-2}	1250
2024	430	1×10^{-2}	525
5083	530	3×10^{-3}	590
A356	530	1×10^{-3}	650
Al–4Mg–1Zr	525	1×10^{-1}	1280
Al–Zn–Mg–Sc	310	3×10^{-2}	1800

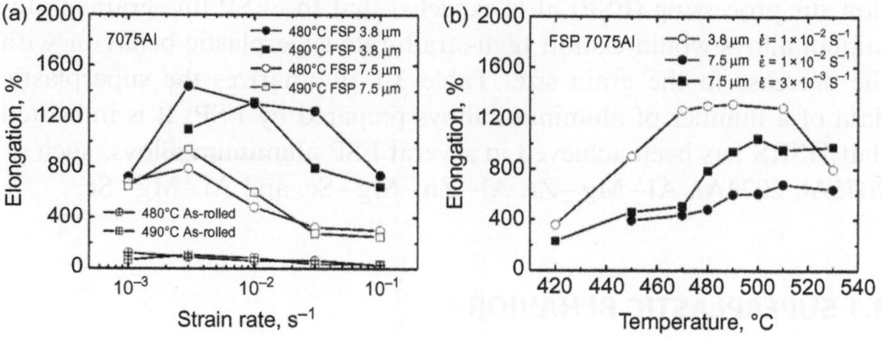

Figure 3.1 Variation of elongation with (a) initial strain rate and (b) test temperature for as-rolled and FSP 7075Al alloys [24].

3.8 μm-7075Al alloys still exhibited large superplastic elongations of 640% and 800%, respectively, for an initial strain rate of 1×10^{-2} s^{-1}.

With decreasing the grain size further, the optimum strain rate for maximum superplastic ductility continues to increase according to the constitutive relationship for superplasticity. Figure 3.2(a) shows the variation of elongation with initial strain rate for FSP fine-grained Al–4Mg–1Zr alloy with an average size of 1.5 μm. Optimum superplasticity was observed at a high strain rate of 1×10^{-1} s^{-1} (even 3×10^{-1} s^{-1} for 450°C) at temperatures ranging from 425°C to 525°C. Maximum superplastic elongation of 1280% was obtained at 525°C and 1×10^{-1} s^{-1}. The optimum strain rate for the FSP Al–4Mg–1Zr is one order of magnitude larger than that for the FSP 3.8 μm-7075Al alloy. Furthermore, the FSP fine-grained Al–4Mg–1Zr alloy exhibited an excellent thermal stability due to the presence of the fine Al_3Zr particles. Even at a higher temperature of 550°C, the FSP Al–4Mg–1Zr exhibited

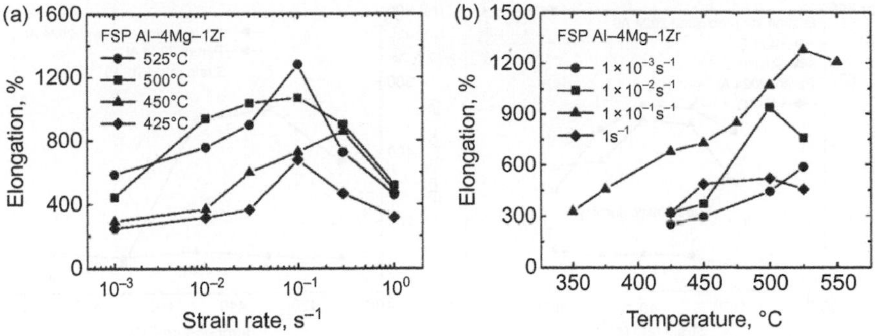

Figure 3.2 Variation of elongation with (a) initial strain rate and (b) test temperature for FSP 1.5 μm Al–4Mg–1Zr [25].

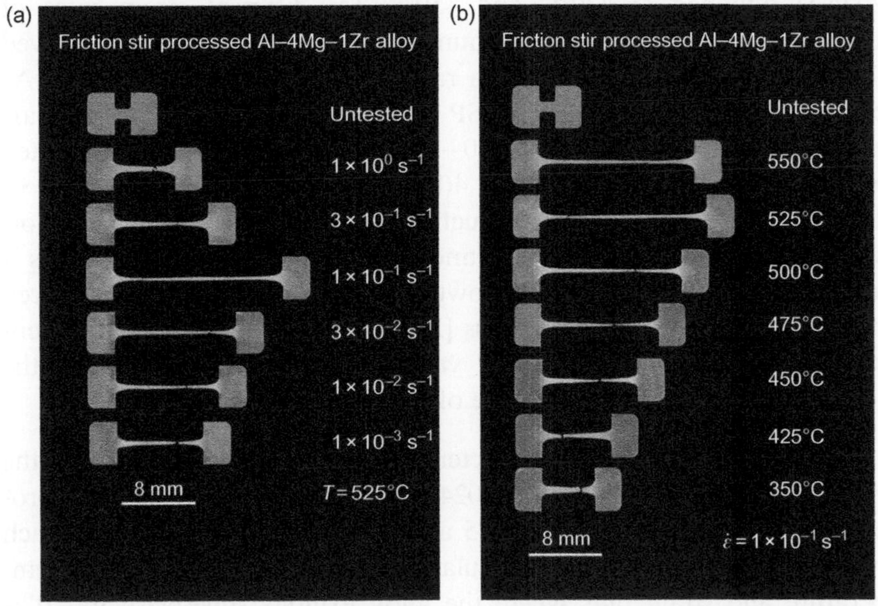

Figure 3.3 Appearance of specimens before and after superplastic deformation at (a) 525°C and different strain rates and (b) 1×10^{-1} s^{-1} and different temperatures [25].

a superplastic elongation of 1210% (Figure 3.2(b)). Figure 3.3 shows the tested specimens of the FSP Al–4Mg–1Zr alloy deformed to failure at 1×10^{-1} s^{-1} for different temperatures. The specimens show neck-free elongation that is characteristic of superplastic flow.

Similarly, FSP 2024Al alloy with an average grain size of 3.9 μm also exhibited HSRS. Figure 3.4(a) presents ductility data of the FSP

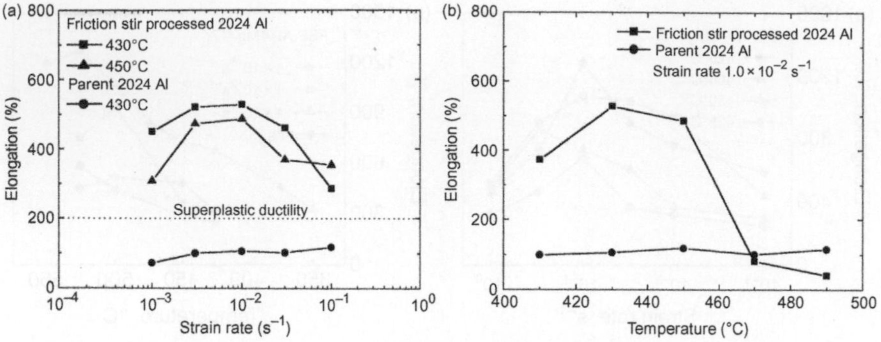

Figure 3.4 Variation of elongation with (a) initial strain rate and (b) test temperature for FSP 3.9 μm-2024Al [29].

2024Al obtained at 430°C and 450°C for different strain rates of 1×10^{-3}–1×10^{-1} s^{-1}. A maximum ductility of ~525% was achieved for FSP 2024Al alloy at a strain rate of 1×10^{-2} s^{-1} and 430°C. As compared to the results of the FSP alloy, the ductility data of the parent alloy are quite marginal (~70–100%). Ductility values are plotted against temperature in Figure 3.4(b) for the optimum strain rate of 1×10^{-2} s^{-1}. It shows that the ductility of FSP 2024Al alloy drops off significantly at and above 470°C and it is only ~41% at 490°C. This is attributed to abnormal grain growth in FSP 2024Al alloy, evidenced by unusual increase in flow stress [29]. On the other hand, the parent material shows modest ductility values (105–120%) all through the temperature range at a strain rate of 1×10^{-2} s^{-1}.

To highlight the HSRS characteristics of FSP aluminum alloys, the relative superplastic domains of 2024Al alloys prepared by different processes are illustrated in Figure 3.5 as strain rate vs. temperature. Each regime is outlined with a rectangular region to show the range of temperature and strain rate, where the alloy exhibits superplasticity. It is noted from Figure 3.5 that conventional thermo-mechanically processed (TMP) 2024Al alloys do not exhibit HSRS [35–37]. Superplasticity in aluminum alloys is typically manifested at ~475°C ($0.8T_m$) or higher temperatures. The optimum superplasticity temperature of 430°C ($0.75T_m$) for the FSP 2024Al is at least ~50°C lower than that for aluminum alloys prepared by the conventional techniques.

It is quite obvious from Figure 3.5 that HSRS can be achieved in the P/M 2024Al (modified composition) [38,39] at a very high strain rate (3 s^{-1}) and higher temperature (500°C). This is attributed to its

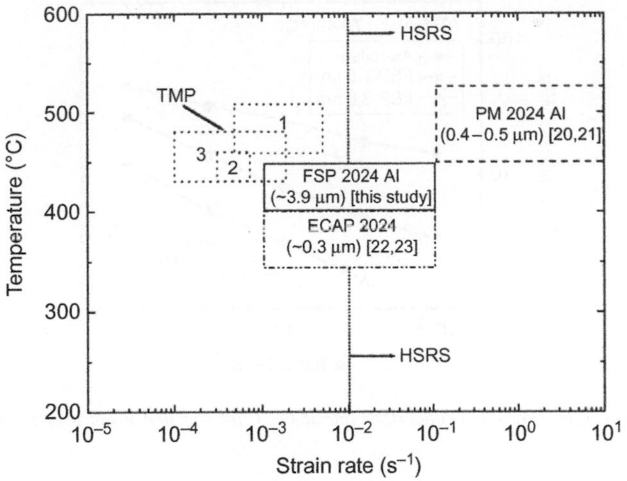

Figure 3.5 A temperature–strain rate map depicting the superplastic domains for 2024 Al alloys processed by different routes [29].

fine-grain microstructure stabilized by intermetallic particles. However, the material processed by powder metallurgy route is expensive and being a modified composition alloy, its knowledge database for other properties is also limited. Although superplastic behaviors of the ECAP 2024Al and the FSP 2024Al alloys are comparable [40,41], optimum superplasticity was achieved at a lower temperature in the ECAP 2024Al primarily due to finer grain size. However, it can be noted that the ECAP 2024Al was pressed for eight passes to achieve optimum superplastic microstructure. For FSP 2024Al, only one pass is needed to produce fine-grained microstructure.

Figure 3.6 shows the effect of FSP on flow stress (at true strain of 0.1) of 7075Al alloys as a function of initial strain rate. FSP resulted in significantly reduced flow stress in 7075Al alloy, which decreased with decreasing grain size. Low flow stress is one of the characteristics of superplastic deformation. Furthermore, the FSP 7075Al exhibited a strain rate sensitivity (m) of $\sim$0.5 within the investigated strain rate range, indicating that grain boundary sliding (GBS) is main deformation mechanism of the FSP 7075Al. Such a high-strain-rate sensitivity value generally indicates higher resistance to neck-free elongation.

Figure 3.7 shows the flow-stress–initial-strain rate curves at different temperatures for the FSP Al–4Mg–1Zr. Different from the FSP 7075Al alloy, FSP Al–4Mg–1Zr exhibited increasing strain rate sensitivity with

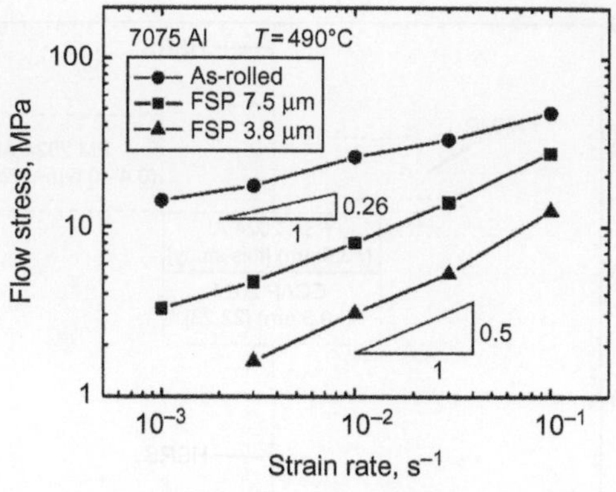

Figure 3.6 Variation of flow stress with initial strain rate at 490° C for FSP and as-rolled 7075Al [24].

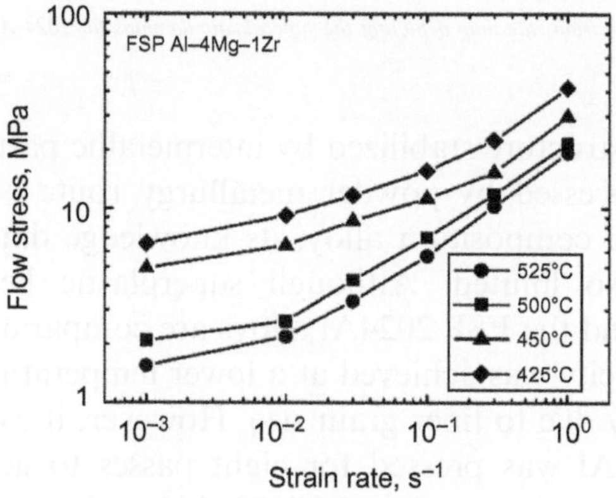

Figure 3.7 Variation of flow stress with initial strain rate for FSP 1.5 μm Al–4Mg–1Zr [26].

increasing initial strain rate from 1×10^{-3} to $1\ s^{-1}$ at the investigated temperatures. For example, at 525°C, the FSP Al–4Mg–1Zr exhibited an *m* value of ~0.15, 0.42, and 0.53 in the initial strain rate ranges of 1×10^{-3}–$1 \times 10^{-2}\ s^{-1}$, 1×10^{-2}–$1 \times 10^{-1}\ s^{-1}$, and 1×10^{-1}–$1\ s^{-1}$, respectively.

3.2 MICROSTRUCTURAL EVOLUTION DURING SUPERPLASTIC DEFORMATION

A relatively stable microstructure is a prerequisite for achieving continuous superplastic deformation. However, during the superplastic deformation, grain growth and elongation are inevitable. It is well accepted that grain growth during superplastic deformation of fine-grained

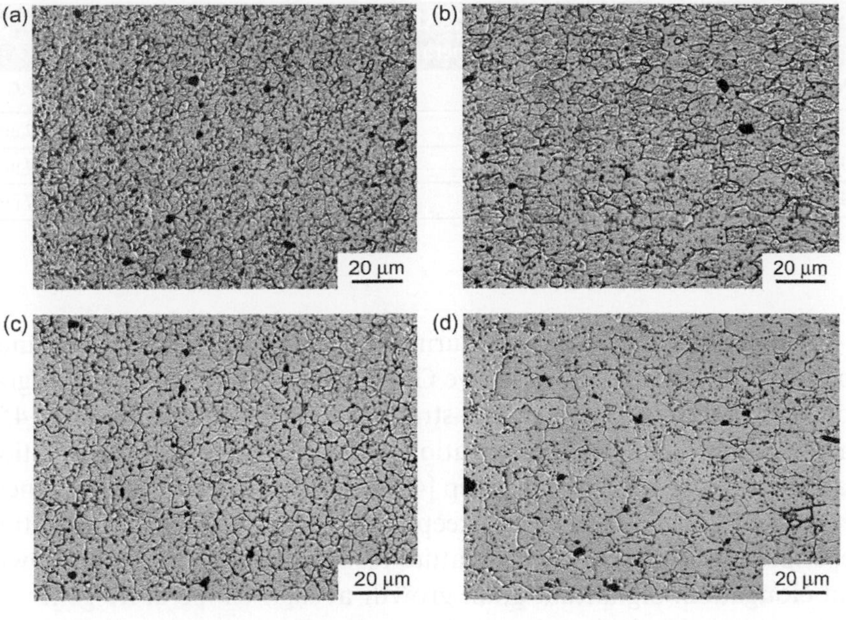

Figure 3.8 Optical micrographs showing grain growth and elongation in FSP 7.8 μm-7075Al deformed to an elongation of 200% at 3×10^{-3} s^{-1} and at (a) grip region, 490°C; (b) gage region, 490°C; (c) grip region, 500°C; and (d) gage region, 500°C. (Tensile axis is horizontal.) [24]

materials is significantly more rapid than static grain growth [42–46]. Recently it was reported [44–46] that noticeable grain elongation along the stress axis occurs in the initial stage of superplastic deformation of fine-grained aluminum alloys.

Figure 3.8 shows the grain structure of the FSP 7.5 μm-7075Al deformed to an elongation of 200% at 3×10^{-3} s^{-1} and different test temperatures of 490°C and 500°C. Compared to the grip region (Figure 3.8(a) and (c) represent static grain growth), grains in the gage region (Figure 3.8(b) and (d)) underwent significant growth and grain shape change. The quantitative measurement of grain size and aspect ratio (Table 3.2) reveals the following two findings. Firstly, significant dynamic grain growth and grain shape change took place in the gage region during superplastic deformation. For example, after deformation to 200% at 500°C and 3×10^{-3} s^{-1}, the 7.5 μm-7075Al exhibited a grain size of 16.4 and 10.7 μm in longitudinal and traverse directions, respectively, and the grain aspect ratio reached 1.5. By comparison, the grain size in the grip region was 8.2 μm. Secondly, both dynamic grain growth and grain shape variation at 500°C were much greater than those at 490°C.

Table 3.2 Grain Size of FSP 7.5 μm-7075Al Deformed to an Elongation of 200% at $3 \times 10^{-3}\ s^{-1}$ and Different Test Temperatures

Position	490°C	500°C
Grip region	7.8 μm	8.2 μm
Gage region parallel to stress axis	13.1 μm	16.4 μm
Gage region traverse to stress axis	9.1 μm	10.7 μm

It is generally believed that during the deformation of fine-grained polycrystalline materials, extensive GBS leads to grain boundary migration (GBM) thereby resulting in strain-induced grain growth [43–45]. On the other hand, grain elongation during superplastic deformation was attributed to dislocation creep [43] or diffusion creep [46]. As these processes (GBM, dislocation creep, and diffusion) are temperature dependent, an increase in temperature results in enhanced grain growth and elongation. Significant grain growth at high temperature generally leads to strain hardening, thereby increasing flow stress.

For rolled or extruded aluminum alloys, it is well documented that the unrecrystallized microstructure gradually evolved into a structure consisting of new recrystallized grains by increasing the strain during superplastic deformation [47–49]. Approximately equiaxed grain structure with nearly random texture and misorientation distribution appeared at the final stage of the superplastic deformation.

Figure 3.9(a) shows the Electron Backscattered Diffraction (EBSD) map of the extruded Al–5.33Mg– 0.23Sc sample. It was indicated that the extruded sample contained a heavily deformed microstructure comprised of banded unrecrystallized grains and fine recrystallized grains along the extruding direction. Figure 3.9(b) shows the frequency distribution of boundary misorientation angles for the statically annealed sample. The average misorientation angle was determined to be 21.5° and the fraction of the high-angle grain boundaries (HAGBs) was only 48%. Distinct ⟨111⟩ fiber texture aligned parallel to the extruding direction, with a maximum intensity of 5.05, was observed (Figure 3.9(c)).

During superplastic deformation, the heavily deformed microstructure of the extruded sample was gradually evolved into equiaxed grains. At a strain of 2.1 (close to failure), the microstructure was

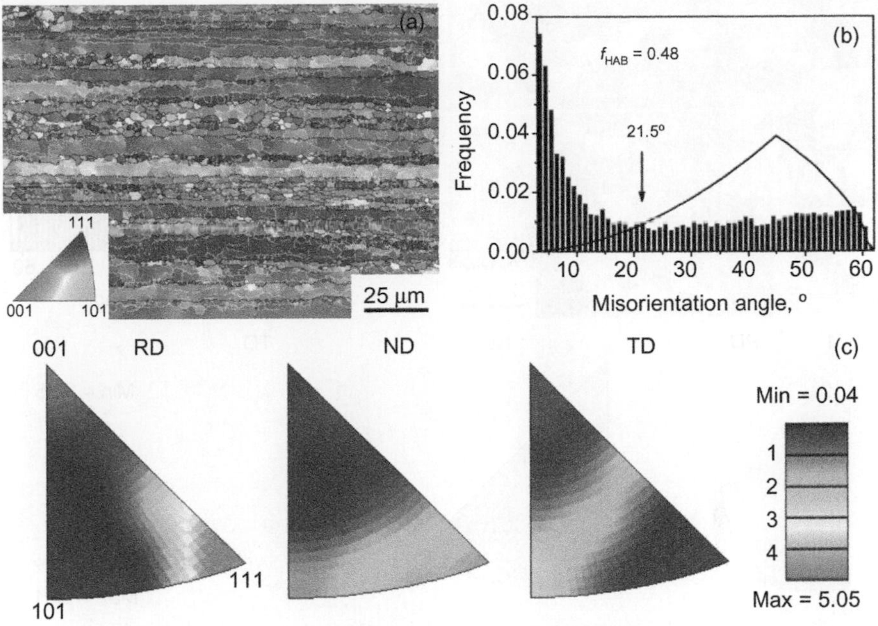

Figure 3.9 Microstructure of extruded Al–5.33Mg–0.23Sc: (a) EBSD map, (b) boundary misorientation angle distribution, and (c) inverse pole figures (RD: extrusion direction; ND: normal direction; TD: transverse direction) [51].

characterized by randomly distributed equiaxed grains with an average grain size of 25.5 μm (Figure 3.10(a)). The misorientation distribution shows a close match with the theoretical distribution (Figure 3.10(b)). The average misorientation angle and the fraction of HAGBs were 37.1° and 94%, respectively, and very close to 40.7° and 97% for the random misorientation distribution predicted by Mackenzie [50]. The maximum orientation density of the ⟨111⟩ fiber texture was only 1.81 (Figure 3.10(c)).

The microstructural evolution of the extruded Al–Mg–Sc could be divided into three stages [51]. Subgrain rotation and coalescence in the early stage resulted in strain hardening. Dynamic recrystallization in the middle stage accounted for the decrease of flow stress. The main deformation mechanism in the final stage was GBS and dynamic grain growth.

Unlike rolled or extruded aluminum alloys, FSP aluminum alloys usually exhibited a fully recrystallized microstructure with uniform and equiaxed grains and predominant HAGBs. As shown in Figure 3.11(a)

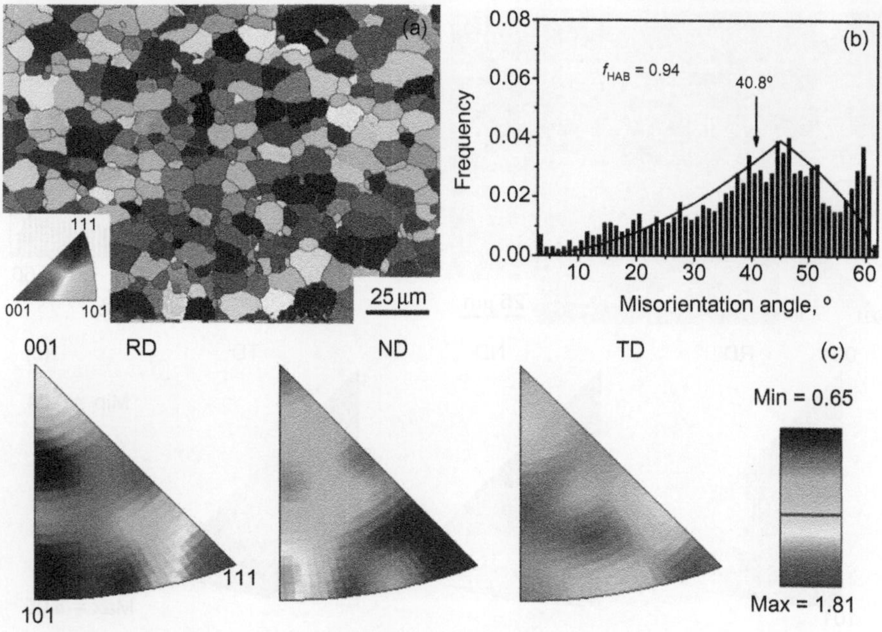

Figure 3.10 Microstructure of extruded Al–5.33Mg–0.23Sc deformed to a strain of 2.1: (a) EBSD map, (b) boundary misorientation angle distribution, and (c) inverse pole figures [51].

and (b), FSP Al–5.33Mg–0.23Sc exhibited fine and equiaxed grain microstructure with an average grain size of ~2.6 μm and an average misorientation angle of 40.3° and an HAGB fraction of 97%, which are very close to the grain assembly for randomly oriented cubes [50]. Furthermore, it was found that the FSP Al–Mg–Sc exhibited a very weak texture component as indicated by the inverse pole figures in Figure 3.11(c).

During superplastic deformation, the grains in the FSP Al–Mg–Sc alloy remained equiaxed and randomly distributed, but the sizes of grains increased. At a strain of 3.1 (close to failure), the FSP Al–Mg–Sc showed a very weak, almost random, texture, whereas the misorientation distributions remained almost unchanged and matched well with the theoretical distribution (Figure 3.12).

The microstructural evolution observed in the FSP Al–Mg–Sc during superplastic deformation is associated with GBS and dynamic grain growth. Such microstructural evolution is schematically shown in Figure 3.13. In the framework of this model, the combined action

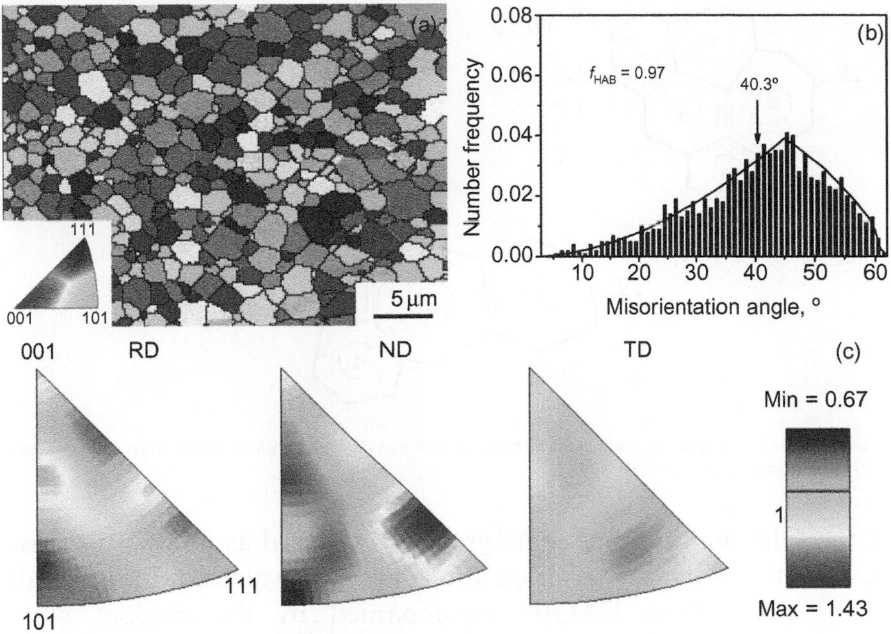

Figure 3.11 Microstructure of FSP Al–5.33Mg–0.23Sc: (a) EBSD map, (b) boundary misorientation angle distribution, and (c) inverse pole figures (RD: FSP direction; ND: normal direction; TD: transverse direction) [51].

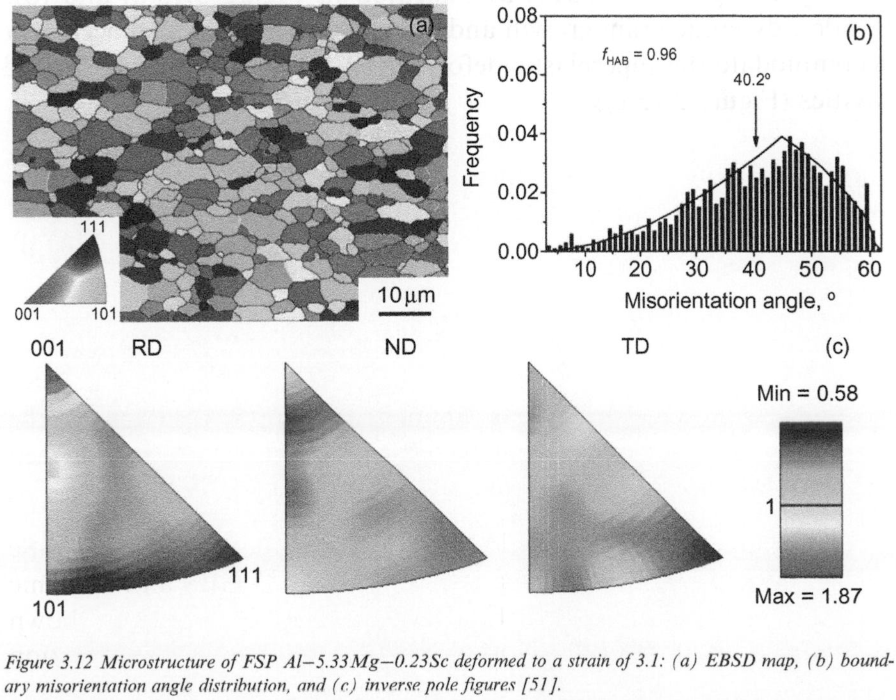

Figure 3.12 Microstructure of FSP Al–5.33Mg–0.23Sc deformed to a strain of 3.1: (a) EBSD map, (b) boundary misorientation angle distribution, and (c) inverse pole figures [51].

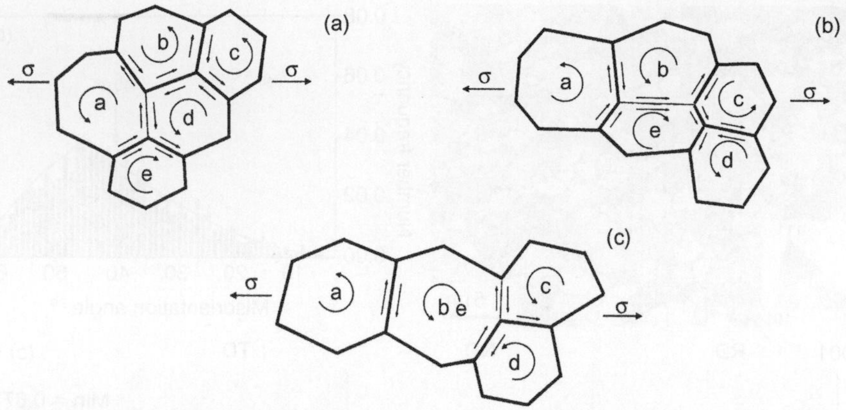

Figure 3.13 Schematic representations of microstructural evolution during superplastic deformation accommodated by GBS and dynamic grain growth [51].

of the GBS and dynamic grain growth is realized as follows. Because most of the boundary misorientations are high values, GBS proceeds easily along these HAGBs accompanied by the random grain rotations during superplastic deformation. Most of low-angle grain boundaries (LAGBs) are transformed to HAGBs due to the GBS and random grain rotations, while parts of HAGBs are simultaneously changed to LAGBs, as indicated in Figure 3.13(a) and (b). Besides, dynamic grain growth and grain shape change also occurs to accommodate the superplastic deformation to avoid the formation of cavities (Figure 3.13(c)).

CHAPTER 4

Low-Temperature Superplasticity

Conventional superplasticity is typically achieved at slow strain rates (10^{-4}–10^{-3} s^{-1}) and higher homologous temperatures ($0.80T_m$, where T_m is the melting point of the matrix alloy expressed in K). It is well established that a decrease in grain size enhances the optimum superplastic response by lowering temperature and/or increasing strain rate. Therefore, if the grain size could be reduced to ultrafine-grained (UFG) range, low-temperature superplasticity (LTSP) could be achieved. LTSP is attractive for commercial superplastic forming, for lowering energy requirement, increasing life for conventional or cheaper forming dies, improving surface quality of the formed component, preventing severe grain growth and solute-loss from the surface layers, thus, resulting in better post-forming mechanical properties [5].

Ideal LTSP should be obtained at $\sim 0.5T_m$ temperature. For aluminum alloys, the corresponding temperature would be about ~190°C or less. Unfortunately, it is difficult to achieve superplasticity at this temperature range even for any UFG Al alloys. For aluminum alloys, the superplasticity achieved at temperatures ≤350°C is usually considered as LTSP. Table 4.1 summarizes the LTSP results of various fine-grained aluminum alloys prepared by different processing approaches [5,10–19]. It is indicated that several plastic working techniques, such as thermo-mechanical processing (TMP), equal channel angular pressing (ECAP), high-pressure torsion (HPT), multiaxial alternative forging (MAF), accumulative roll bonding (ARB), have been used to produce fine-grained aluminum alloys of 0.1–1.2 μm that exhibit LTSP. As presented above, friction stir processing (FSP) can produce an UFG structure of down to ~0.1 μm [52]. Therefore, it is very likely that FSP UFG aluminum alloys will exhibit LTSP behavior.

4.1 LTSP OF FSP AL–ZN–MG–SC

An ingot casting of Al–8.87Zn–2.56Mg–0.091Sc (in wt.%) was subjected to FSP at a tool rotation rate of 400 rpm and a traverse speed of 25.4 mm/min. Figure 4.1(a) shows an optical macrograph of the

Table 4.1 A Comparison of Low-Temperature Superplastic Properties of Various Aluminum Alloys

Alloy	Processing	Grain Size (μm)	Elongation (%)	Temperature (°C)	Strain Rate (s^{-1})	*m* Value	Reference
1420Al	TS (ε 6.0)	~0.1	775 330	300 250	1×10^{-1} 1×10^{-1}	0.38 0.29	[17]
1420Al	ECAP (12 passes)	1.2	>1180 ~620	350 250	1×10^{-2} 1×10^{-3}	–	[16]
Al–6Cu–0.4Zr	ECAP (12 passes)	~0.5	970	300	1×10^{-2}	–	[14]
Al–3Mg–0.2Sc	ECAP (8 passes)	0.2	420	200	3×10^{-4}	–	[15]
Al–3Mg–0.2Sc	ECAP (8 passes)	0.2	1280 ~1700	300 350	1×10^{-2} 3.3×10^{-2}	~0.5 ~0.5	[12]
5083Al	ECAP (8 passes)	0.3	315	275	5×10^{-4}	0.4	[13]
Al–10Mg–0.1Zr	TMP (12 rolling pass)	–	~1100	300	1×10^{-3}	~0.5	[10]
5083Al	TMP	~0.5	511	230	2×10^{-3}	~0.35	[11]
8090Al	TMP	0.7	710	350	8×10^{-4}	~0.37	[5]
5083Al	MAF (10 cycles)	~0.8	340	200	2.8×10^{-3}	0.39	[18]
5083Al	ARB (5 cycles)	0.28	230	200	1.7×10^{-3}	0.37	[19]
1570Al	ECAP (ε 16.0)	~1.0	~1110	350	1.4×10^{-2}	~0.39 (max)	[56]

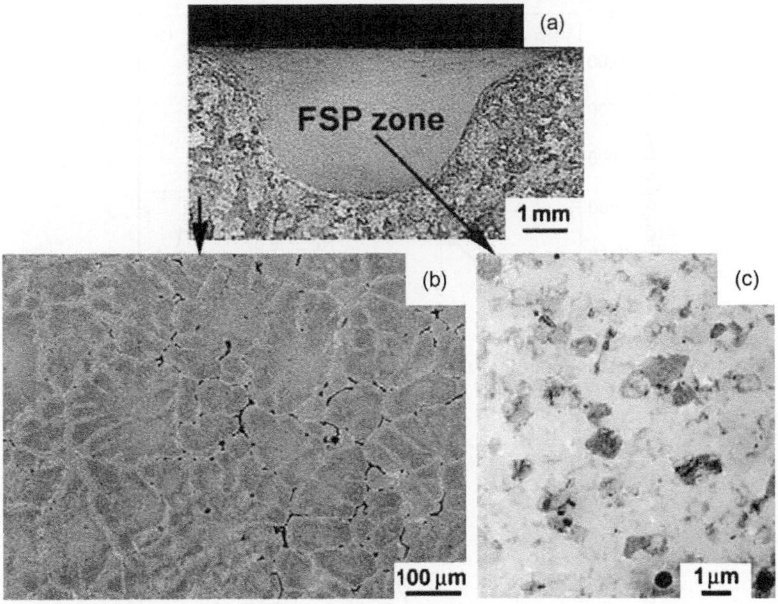

Figure 4.1 Micrographs showing (a) cross-sectional macroscopic view of FSP Al–Zn–Mg–Sc (OM), (b) typical dendritic microstructure in parent zone (OM), and (c) ultrafine grains in nugget zone (TEM) [31].

processed region of the FSP alloy along with the unprocessed zone in the transverse cross-section. As-cast Al–Zn–Mg–Sc alloy exhibited a typical cast structure with dendrites of 200–300 μm spacing, microsegregation and porosity in the interdendritic regions (Figure 4.1(b)). After FSP, a fine and uniform microstructure was achieved with an average grain size of 0.68 μm (Figure 4.1(c)).

Figure 4.2(a) shows the variation of elongation with testing temperature at an initial strain rate of $1 \times 10^{-2}\,s^{-1}$ for the FSP Al–Zn–Mg–Sc alloy. It shows that superplasticity was achieved in a wide range of low temperatures (220–390°C) and an optimum ductility of 1165% was obtained at a strain rate of $3 \times 10^{-2}\,s^{-1}$ and 310°C. It is worth mentioning that an elongation of 266% was attained at a low temperature of 220°C and at a high strain rate of $1 \times 10^{-2}\,s^{-1}$. Figure 4.2(b) shows the effect of initial strain rate on elongation at different temperatures. A few interesting observations can be made from this figure. Firstly, at a low temperature of 220°C, a maximum ductility of 525% was obtained at an initial strain rate of $1 \times 10^{-3}\,s^{-1}$. Secondly, at higher temperatures (250°C onward), the optimum strain rate increased by more than an order of magnitude. The optimum ductility of 1165% was obtained at a strain rate of $3 \times 10^{-2}\,s^{-1}$ and

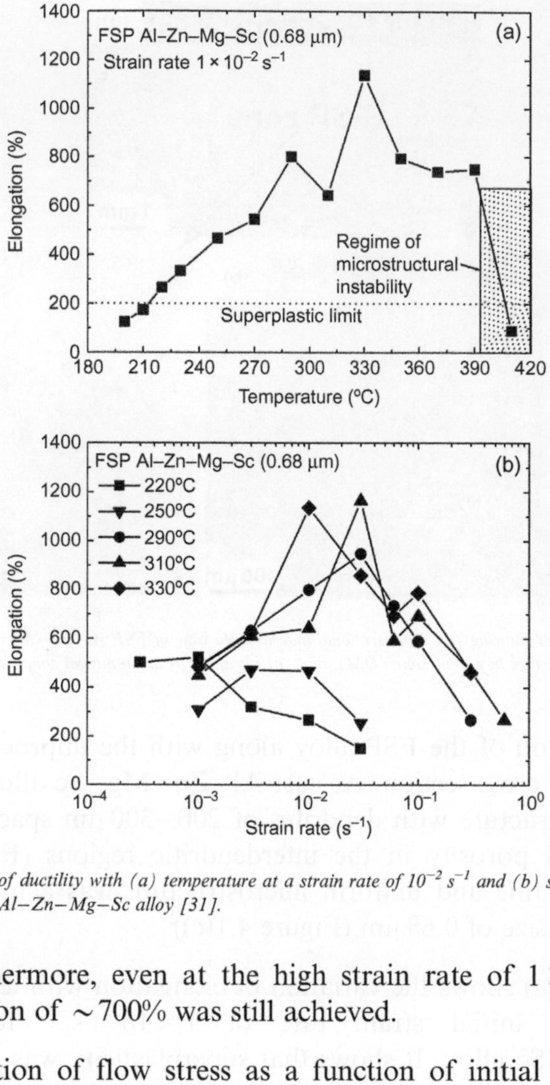

Figure 4.2 Variation of ductility with (a) temperature at a strain rate of 10^{-2} s^{-1} and (b) strain rate at different temperatures in FSP Al–Zn–Mg–Sc alloy [31].

310°C. Furthermore, even at the high strain rate of 1×10^{-1} s^{-1}, a high elongation of ~700% was still achieved.

The variation of flow stress as a function of initial strain rate is shown on double logarithmic scale in Figure 4.3 in the temperature range of 220–330°C. The apparent *m* values (slopes of the fitted straight lines) were 0.33 ± 0.3. The activation energy was calculated to be ~142 kJ/mol [31]. This value is identical to the activation energy value for lattice self-diffusion in Al [53].

4.2 LTSP OF FSP AL–MG–ZR

As given in Table 4.2, LTSP could be achieved in fine-grained aluminum alloys produced by plastic working techniques, such as

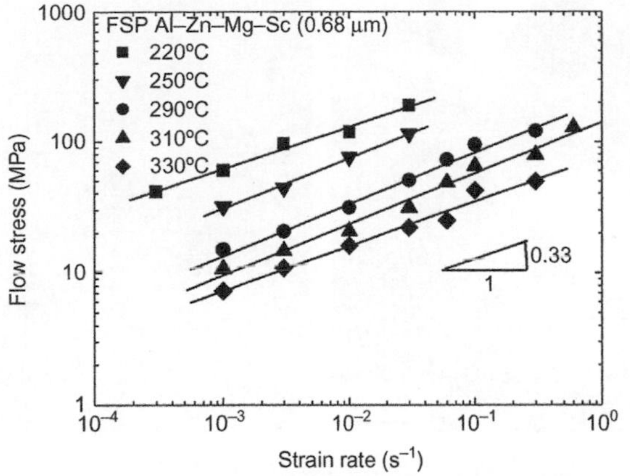

Figure 4.3 Variation of flow stress with strain rates at different testing temperatures in FSP Al–Zn–Mg–Sc alloy [31].

torsional straining (TS), ECAP, MAF, ARB, and TMP. However, previous investigations on LTSP were limited to temperatures of ≥200°C ($0.51T_m$). No studies on LTSP of aluminum alloys at temperatures of $<0.5T_m$ have been reported. A scientific curiosity is whether superplasticity can be developed in fine-grained aluminum alloys at temperatures $<0.50T_m$. An UFG FSP aluminum alloy with high fraction of high-angle grain boundaries (HAGBs) is an ideal microstructure to check low-temperature limit for manifestation of superplasticity.

Figure 4.4 shows the microstructures of the as-extruded and FSP Al–4Mg–1Zr alloys. For the extruded sample, the microstructure was characterized by predominantly low-angle grain boundaries with grains/subgrains aligned along the extrusion direction, and the grain size was nonuniform and the average subgrain size was 1.8 µm (Figure 4.4(a)). For the FSP samples (Figure 4.4(b) and (c)), the microstructures were characterized by uniform and equiaxed recrystallized grains with predominantly HAGBs. The average grain size in the FSP sample prepared by the standard tool was 1.6 µm (hereafter referred as micron-grained FSP sample). However, when a modified tool with reduced shoulder and pin diameters was used, the average grain size was reduced to ~0.7 µm (hereafter referred as UFG FSP sample).

The frequency distribution of boundary misorientation angles for the UFG FSP sample is shown in Figure 4.5. The fraction of HAGBs was measured to be 97%. This is similar to that observed in micron-FSP aluminum alloys (Figure 2.2). For comparison, the theoretical

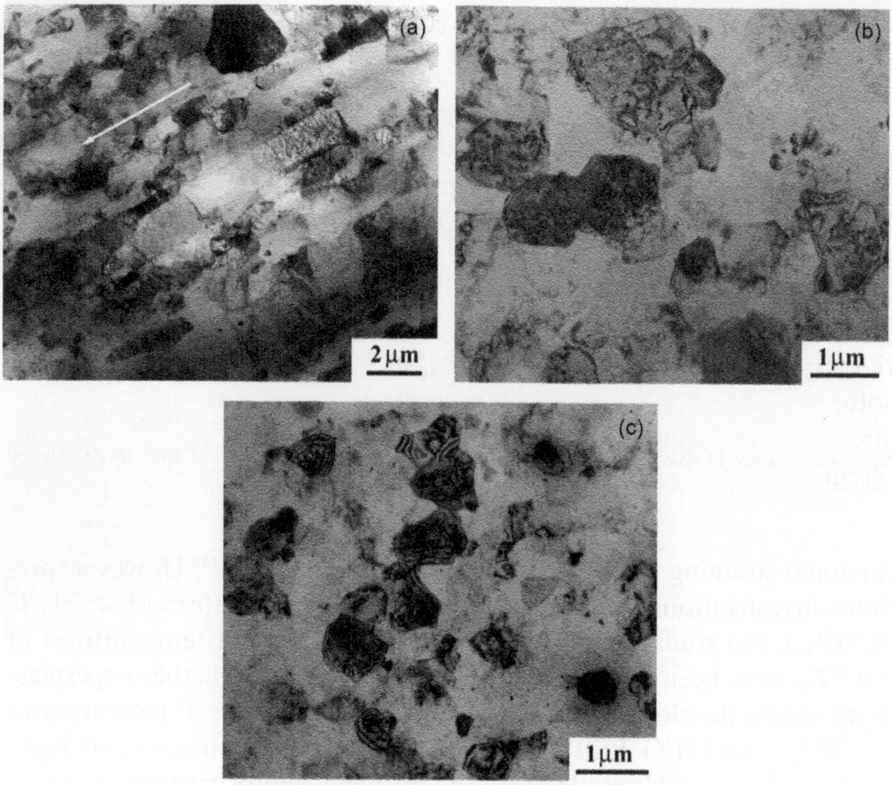

Figure 4.4 TEM micrographs of Al–4Mg–1Zr: (a) as-extruded (the arrow denotes the extrusion direction), (b) FSP with a grain size of 1.6 μm, and (c) FSP with a grain size of 0.7 μm (the FSP direction is vertical to the page) [27].

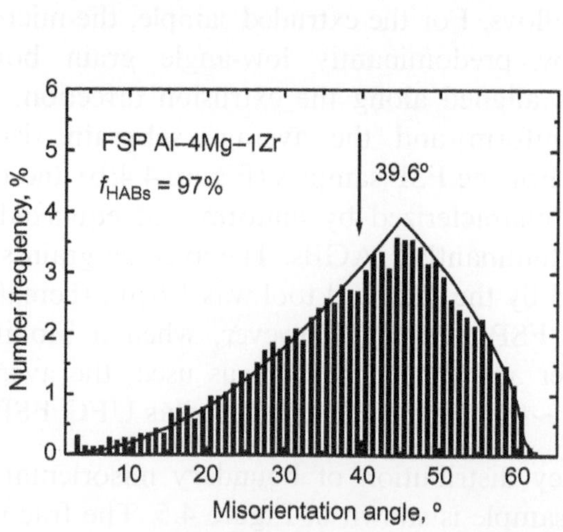

Figure 4.5 Grain boundary misorientation angle distribution for 0.7 μm FSP Al–4Mg–1Zr [142].

distribution of grain boundary misorientation angles for a random polycrystal of cubic structure [50] is also shown in Figure 4.5 by a black solid line. It is seen that the misorientation distribution for the FSP Al–4Mg–1Zr shows a close match with the theoretical distribution.

The stress–strain behavior of the UFG FSP sample is shown in Figure 4.6(a) as a function of initial strain rate at 175°C. The optimum strain rate for maximum elongation was $1 \times 10^{-4}\ s^{-1}$. The UFG FSP sample exhibited a well-behaved stress–strain curve, where the flow stress remained almost constant during the superplastic flow, at the initial strain rates of 1×10^{-4}–$3 \times 10^{-3}\ s^{-1}$, whereas a continuous strain softening was observed at a lower strain rate of $5 \times 10^{-5}\ s^{-1}$. The stress–strain behavior of the UFG FSP sample at 175°C is quite different from that of micron-grained FSP alloys at higher temperatures

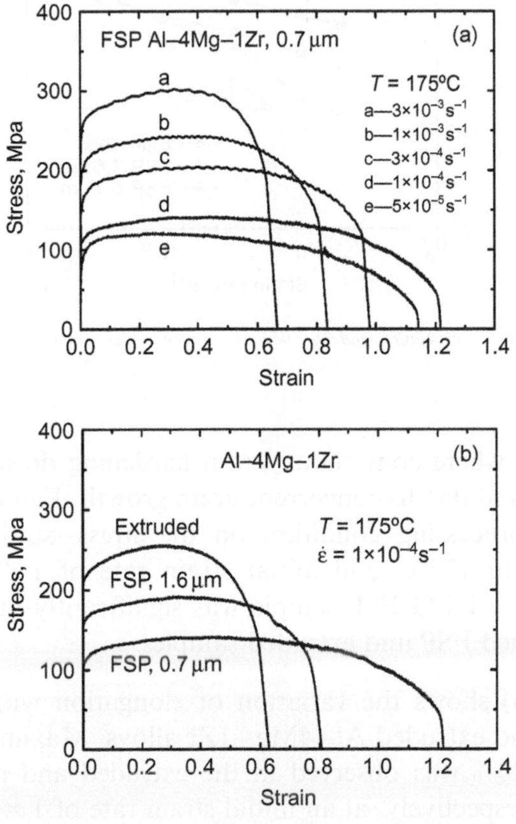

Figure 4.6 (a) Stress–strain behavior of UFG FSP Al–4Mg–1Zr as a function of initial strain rate at 175°C and (b) effect of processing condition on stress–strain behavior of Al–4Mg–1Zr at 175°C and an initial strain rate of $1 \times 10^{-4}\ s^{-1}$ [27].

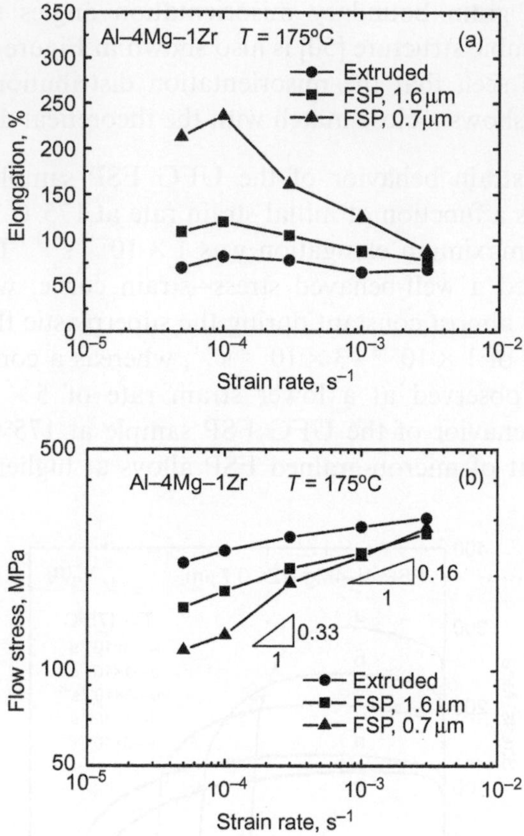

Figure 4.7 Variation of (a) elongation and (b) flow stress with initial strain rate for extruded and FSP Al–4Mg–1Zr [27].

of 425–570°C, where continuous strain hardening during superplastic flow was observed due to concurrent grain growth. Figure 4.6(b) shows the effect of processing condition on the stress–strain behavior of Al–4Mg–1Zr at 175°C and initial strain rate of 1×10^{-4} s^{-1}. The elongation of the UFG FSP sample was significantly higher than that of micron-grained FSP and extruded samples.

Figure 4.7(a) shows the variation of elongation with initial strain rate for FSP and extruded Al–4Mg–1Zr alloys. Maximum elongation of 82% and 120% was observed in the extruded and micron-grained FSP samples, respectively, at an initial strain rate of 1×10^{-4} s^{-1}. This means that both extruded and micron-grained FSP samples did not exhibit superplasticity at a low temperature of 175°C. For the UFG

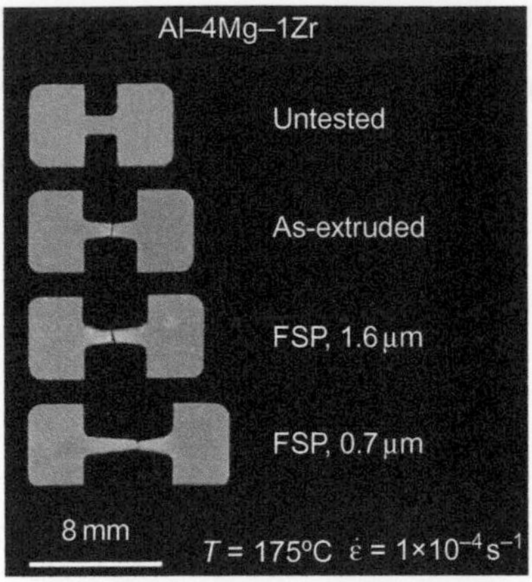

Figure 4.8 Appearance of specimens before and after superplastic deformation at 175° C and an initial strain rate of $1 \times 10^{-4}\ s^{-1}$ [27].

FSP sample, a maximum elongation of 240% was attained at an initial strain rate of $1 \times 10^{-4}\ s^{-1}$ for 175°C. This indicates that LTSP at 175°C is developed in the UFG FSP Al–4Mg–1Zr alloy.

Figure 4.7(b) shows the variation of flow stress with initial strain rate for FSP and extruded Al–4Mg–1Zr alloys. The strain rate sensitivity of both extruded and micron-grained FSP samples were consistently lower than ~0.16 throughout the investigated strain rates of 5×10^{-5}–$3 \times 10^{-3}\ s^{-1}$. This accounts for the absence of superplasticity in the extruded and micron-grained FSP samples. By comparison, the strain rate sensitivity of the UFG FSP sample is 0.16, 0.34, and 0.17 in the initial strain rate ranges of 5×10^{-5}–$1 \times 10^{-4}\ s^{-1}$, 1×10^{-4}–$3 \times 10^{-4}\ s^{-1}$, and 3×10^{-4}–$3 \times 10^{-3}\ s^{-1}$, respectively, i.e., the UFG FSP sample exhibited the typical S-type stress–strain rate behavior characteristic of a superplastic material. The maximum strain rate sensitivity of 0.34 at the initial strain rate of $1 \times 10^{-4}\ s^{-1}$ corresponds to the maximum elongation of 240%. Furthermore, flow stress of the UFG FSP sample is significantly lower than that of the extruded and micron-grained samples at the initial strain rates of 5×10^{-5}–$1 \times 10^{-3}\ s^{-1}$. This is attributed to a significantly refined microstructure.

Figure 4.8 shows tested Al–4Mg–1Zr specimens deformed to failure at 175°C and an initial strain rate of $1 \times 10^{-4}\ s^{-1}$. The UFG FSP specimen shows relatively uniform elongation characteristic of superplastic flow.

Table 4.2 A Comparison of Low-Temperature Superplastic Properties of Various Aluminum Alloys at 200°C

Alloy	Processing	Grain Size (μm)	Elongation (%)	Temperature (°C)	Strain Rate (s^{-1})	*m* Value	Reference
Al–3Mg–0.2Sc	ECAP (8 passes)	0.2	420	200	3×10^{-4}	–	[15]
5083Al	MAF (10 cycles)	~0.8	340	200	2.8×10^{-3}	0.39	[18]
5083Al	ARB (5 cycles)	0.28	230	200	1.7×10^{-3}	0.37	[19]
Al–4Mg–1Zr	FSP	0.7	240	175	1×10^{-3}	0.34	[26]

The lowest temperature for LTSP of aluminum alloys previously reported in literature is 200°C [15,18,19]. The processing condition, grain size, and superplastic properties of UFG aluminum alloys at 200°C are summarized in Table 4.2. For comparison, superplastic data of the UFG Al–4Mg–1Zr at 175°C is also included in Table 4.2.

The results show that the simple application of FSP with lower heat-input tool design can induce UFG microstructure and LTSP in an aluminum alloy at $0.48T_m$. These are the first results illustrating superplastic behavior of aluminum alloys at temperatures of $<0.5T_m$. This demonstrates that FSP is a very effective processing technique to create very fine-grained microstructure in aluminum alloys capable of exhibiting high strain rate (HSRS)/LTSP.

Based on superplastic study at 175°C, the UFG Al–4Mg–1Zr sample was subjected to superplastic investigation in a wide temperature range from 175°C to 425°C. Figure 4.9(a) and (b) shows the variation of superplasticity and flow stress with initial strain rate for UFG Al–4Mg–1Zr. It was indicated that the UFG Al–4Mg–1Zr sample exhibited excellent superplasticity at testing temperatures from 175°C to 350°C (Figure 4.9(a)). With increasing the testing temperature from 175°C to 350°C, the optimum strain rate for the maximum superplasticity increased from $1 \times 10^{-4}\ s^{-1}$ to $1 \times 10^{-1}\ s^{-1}$. Furthermore, the maximum superplasticity increased from 240% at 175°C to ~1200% at 275°C and then kept stable superplastic value of ~1200% with further increasing temperature from 275°C to 350°C.

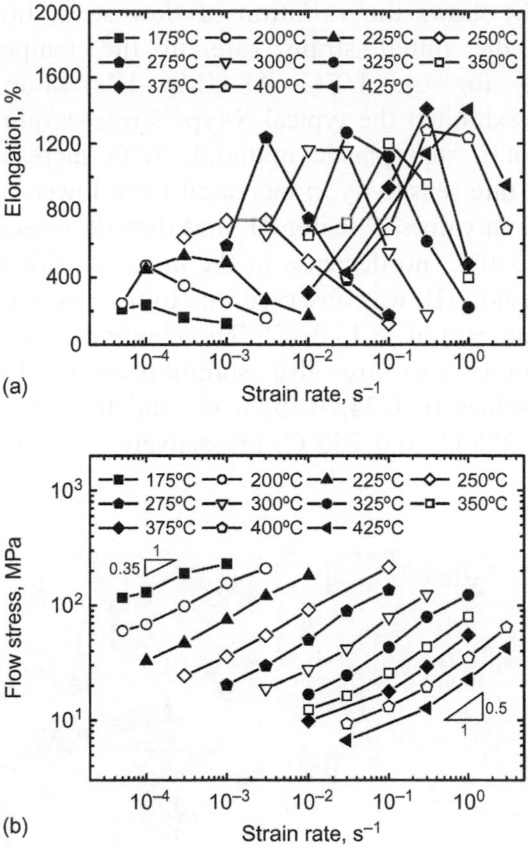

Figure 4.9 Variation of (a) ductility and (b) flow stress with initial strain rate with testing temperatures for FSP 0.7 μm Al–4Mg–1Zr [142].

It is important to note that with increasing testing temperature to high temperature range (375–425°C), the UFG Al–4Mg–1Zr sample exhibited further increased optimum strain rate and maximum superplasticity. At 425°C, the UFG Al–4Mg–1Zr sample exhibited an elongation of ~1400% at strain rate as high as 1 s^{-1} (Figure 4.9(a)). This is attributed to excellent thermal stability of the UFG Al–4Mg–1Zr sample. As shown in Figure 4.10, even after static annealing at 425°C for 20 min, the UFG Al–4Mg–1Zr sample still maintained fine and equiaxed microstructure with a grain size of 0.98 μm. Figure 4.11 clearly shows that the grain size of the UFG Al–4Mg–1Zr sample was fundamentally stable with annealing temperature up to 425°C.

Figure 4.9(b) shows the variation of flow stress (at a true strain of 0.1) with the initial strain rate in the temperature ranges of 175–425°C for the UFG Al–4Mg–1Zr sample. The FSP Al–4Mg–1Zr exhibited the typical S-type stress–strain rate behavior characteristic of a superplastic material. With increasing the strain rate, the strain rate sensitivity *m* increased from lower values in region I to the maximum values in region II, and then decreased in region III. At and above 350°C, no decrease in the *m* value with the strain rate, i.e., a lack in region III, was observed due to the absence of experimental data at strain rate of $\geq 1-3\ s^{-1}$. The maximum *m* values and corresponding strain rates are also summarized in Table 4.3. The maximum *m* values of 0.34, 0.36, 0.41, and 0.41 were observed at 175°C, 200°C, 225°C, and 250°C, respectively. At and above 275°C,

Figure 4.10 Typical microstructures for FSP 0.7 μm Al–4Mg–1Zr after static annealing at (a) 175°C and (b) 425°C [142].

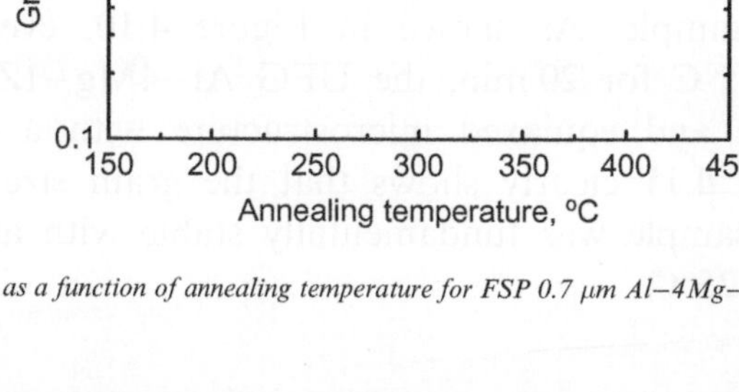

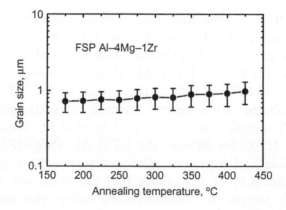

Figure 4.11 Grain size as a function of annealing temperature for FSP 0.7 μm Al–4Mg–1Zr [142].

the maximum m values of ~0.5 were consistently observed. It is noted that the strain rates for the maximum ductility are consistent with those for the maximum m values.

The constitutive equation for superplasticity of fine-grained aluminum alloys predicts that the increase in the testing temperature resulted in an increase in the strain rate if the grain size kept constant. However, for most fine-grained aluminum alloys, such an increasing trend was rarely observed in a wide range of the temperatures due to the remarkable grain growth at high temperatures. As shown in Figure 4.10, the FSP UFG Al–4Mg–1Zr exhibited excellent thermal stability at high temperatures. Therefore it is possible to identify the variation trend of optimum strain rate with the grain size using the FSP UFG Al–4Mg–1Zr.

Figure 4.12(a) shows the variation of superplasticity with the testing temperature at various strain rates. It is noted that the optimum strain rate tended to increase with increasing the temperature. In Figure 4.12(b), the optimum strain rate for maximum superplasticity is plotted as a function of temperature for the FSP UFG Al–4Mg–1Zr. For the purpose of comparison, the superplastic data from other studies [31,54–57] are also included in this plot. It is clear that the optimum strain rate

Table 4.3 A Summary of Superplastic Properties of FSP 0.7 μm Al–4Mg–1Zr Sample at Various Temperatures [142]

Temperature, °C	Optimum Strain Rate, s^{-1}	Maximum Elongation, %	Maximum m Value	Strain Rate Range for Maximum m Value, s^{-1}
175	1×10^{-4}	240	0.34	1×10^{-4}–3×10^{-4}
200	1×10^{-4}	470	0.36	1×10^{-4}–1×10^{-3}
225	3×10^{-4}	530	0.41	3×10^{-4}–3×10^{-3}
250	1×10^{-3}–3×10^{-3}	740	0.41	1×10^{-3}–1×10^{-2}
275	3×10^{-3}	1235	0.48	3×10^{-3}–3×10^{-2}
300	1×10^{-2}–3×10^{-3}	1160	0.53	3×10^{-2}–1×10^{-1}
325	3×10^{-2}	1265	0.51	3×10^{-2}–3×10^{-1}
350	1×10^{-1}	1200	0.49	1×10^{-1}–1
375	3×10^{-1}	1410	0.52	1×10^{-1}–1
400	3×10^{-1}	1280	0.52	3×10^{-1}–3
425	1	1405	0.52	3×10^{-1}–3

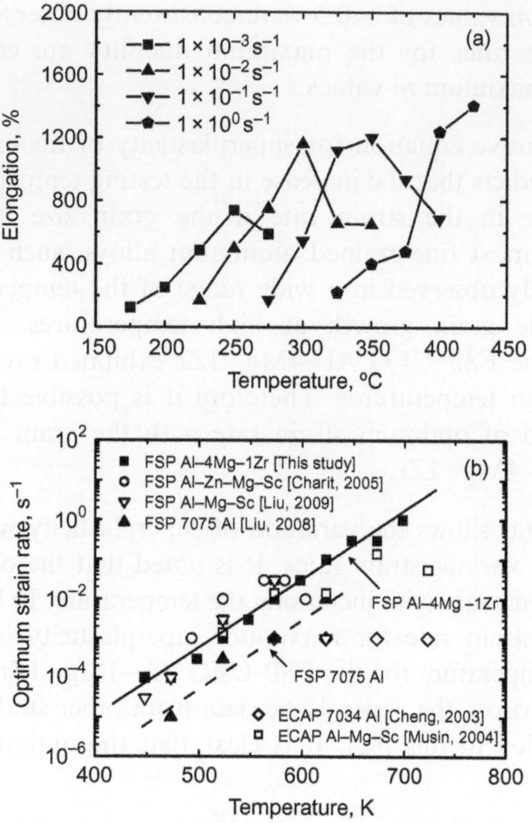

Figure 4.12 (a) Variation of elongation with temperature for various strain rates and (b) variation of optimum strain rate for maximum superplasticity with temperatures for FSP 0.7 μm Al–4Mg–1Zr and other UFG aluminum alloys [142].

($\dot{\varepsilon}_{opti}$) increased with increasing the temperature (T) for the FSP UFG Al–4Mg–1Zr, which can be described as

$$\log \dot{\varepsilon}_{opti} = 0.0182T - 12.36 \tag{4.1}$$

This indicates that a linear relationship exists between $\log \dot{\varepsilon}_{opti}$ and T. For the FSP UFG Al–4Mg–1Zr, the actual grain sizes before deformation remained almost unchanged in the temperature range of 175–425°C due to the effective pinning of numerous Al_3Zr particles on the grain boundaries (Figure 4.10); therefore, the linear variation trend of $\log \dot{\varepsilon}_{opti}$ with T indicates a real effect of the temperature on the optimum strain rate. However, such a variation trend has not been reported in previous studies. This is due to the fact that for most fine-grained aluminum alloys, the grains exhibited an obvious coarsening with increasing the temperature. At high temperatures, the actual grain sizes before superplastic deformation were much larger than the initial ones, which resulted in $\log \dot{\varepsilon}_{\text{opti}}$ deviating from the linear relationship with T. Figure 4.12(b) clearly indicates that the superplastic data from other fine-grained aluminum alloys prepared by FSP or ECAP depart from the linear relationship at higher temperatures.

A similar relationship is also observed in Figure 4.12(b), indicated by a dashed line, for FSP 7075Al whose superplastic data and microstructural evolution have been previously reported [55]. The equation is

$$\log \dot{\varepsilon}_{\text{opti}} = 0.0182T - 13.36 \tag{4.2}$$

This implies that the most possible relationship between $\log \dot{\varepsilon}_{\text{opti}}$ and T for superplasticity of the fine-grained aluminum alloys can be expressed as

$$\log \dot{\varepsilon}_{\text{opti}} = 0.0182T + B \tag{4.3}$$

where B is a constant depending on the material, grain size, and grain boundary structure. It was well documented that reducing the grain size resulted in an increase in the optimum strain rate [24,47,56]. Furthermore, it was suggested that the increase in the fraction of the HAGB was also beneficial to the increase in the optimum strain rate [24,29,54].

Figure 4.12(b) indicates that for a constant temperature, the optimum strain rate for the FSP Al–4Mg–1Zr is one order of magnitude higher than that for the FSP 7075Al, whereas for a constant optimum strain rate, the temperature for the former is 55°C lower than that for the latter. This implies that it is easier to achieve HSRS and/or LTSP in the FSP Al–4Mg–1Zr than in the FSP 7075Al. Because the grain size usually exhibits a remarkable increase with increasing temperature for most of fine-grained aluminum alloys, the data that fit Eq. (4.3) are quite limited. Therefore, the validity and applicability of Eq. (4.3) need more data to support. Furthermore, a further analysis is required to understand the physical meaning and implication of the slope of 0.0182 between $\log \dot{\varepsilon}_{\text{opti}}$ and T.

CHAPTER 5

Superplasticity of Cast Alloy—An Example

A356 is one of the most widely applied commercial Al–Si–Mg alloys, particularly in the aircraft and automotive industries because it has good castability [58] and can be strengthened by artificial aging [59–61]. The microstructure of the cast A356 consists of primary α-aluminum dendrites and interdendritic irregular Al–Si eutectic regions (Figure 5.1(a)). Such a microstructure does not exhibit superplastic behavior due to coarse dendrites and heterogeneous Si particle distribution. Friction stir processing (FSP) generates an intense breakup of the as-cast microstructure and subsequent material mixing, thereby creating a microstructure with fine and equiaxed Si particles uniformly distributed in the aluminum matrix (Figure 5.1(b)). The grain size in the FSP A356 is on the order of ~3 μm (Figure 5.1(c)). This is in good agreement with that obtained in other FSP aluminum alloys [21]. The microstructure of FSP A356 is very similar to that of metal matrix composites. It has been reported that metal matrix composites exhibit superplastic behavior under special conditions [62–64]. Therefore, it is very likely that FSP A356 will exhibit superplastic behavior.

Figure 5.2 shows the variation of elongation as a function of strain rate and temperature for both the FSP and cast conditions. Elongation of the cast A356 was low (<200%) and did not exhibit any appreciable dependence on strain rate or temperature. By comparison, FSP A356 exhibited a maximum elongation of 650% and demonstrated a strain rate and temperature sensitivity with optimum test parameters of 530°C at an initial strain rate of $1 \times 10^{-3}\ s^{-1}$.

Recently, Kim et al. [65] investigated high-temperature deformation behavior of equal channel angular pressed (ECAP) A356. It was reported that an ~2–3 μm fine-grained structure was produced in A356 after six-pass ECAP. However, no superplasticity was obtained in the ECAP A356 with a maximum elongation of ~100% and a strain rate sensitivity of ~0.2 in the investigated strain rate range. This result was attributed to the occurrence of grain growth by Kim et al. [65].

For the FSP A356, it appears that the fine-grained structure is stable at higher temperature. This is supported by the fact that even at the highest temperature of 570°C, FSP A356 still exhibited a relatively high elongation of 515% (Figure 5.2(b)). This may be attributable to

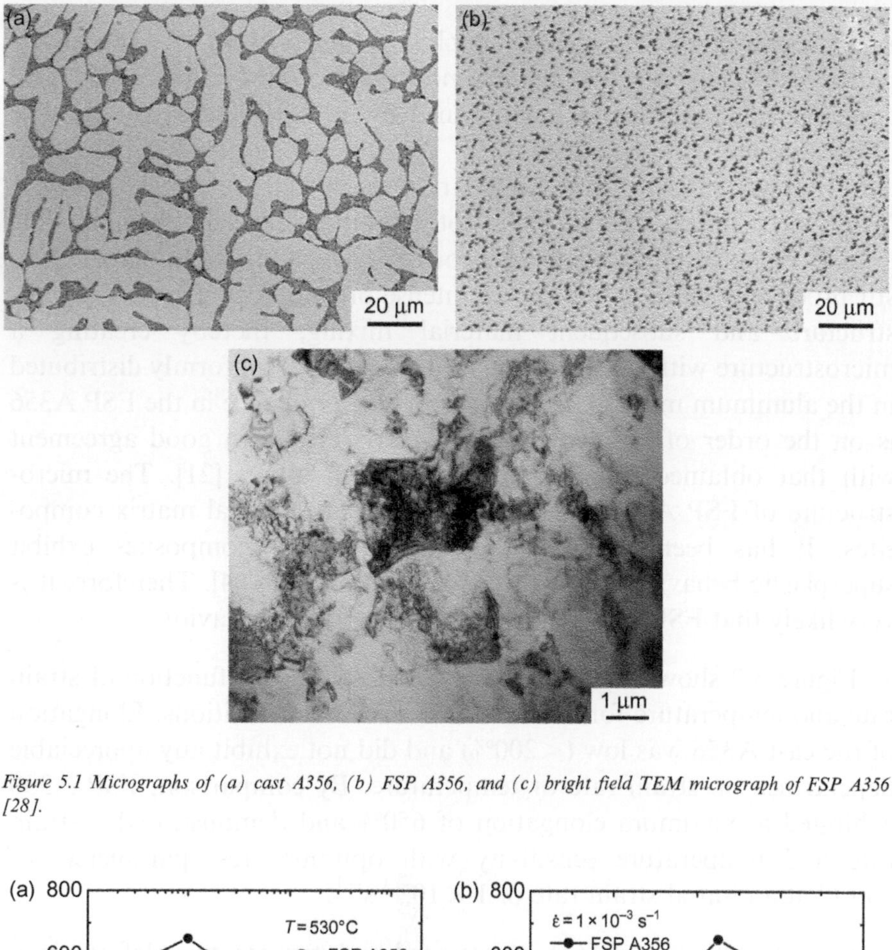

Figure 5.1 Micrographs of (a) cast A356, (b) FSP A356, and (c) bright field TEM micrograph of FSP A356 [28].

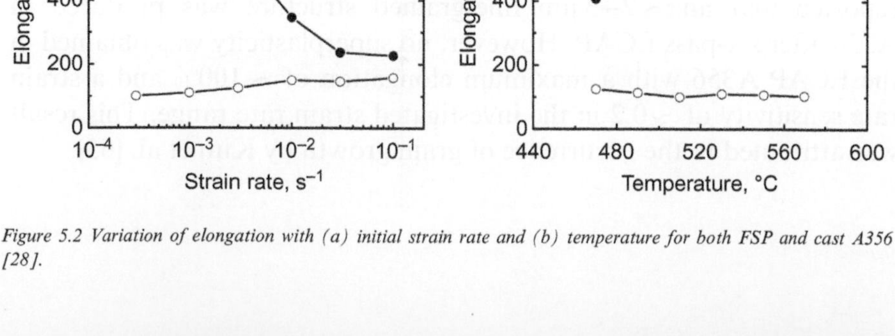

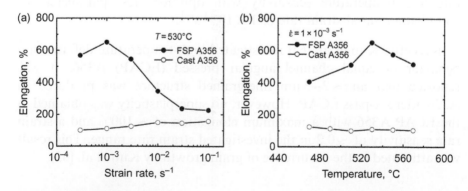

Figure 5.2 Variation of elongation with (a) initial strain rate and (b) temperature for both FSP and cast A356 [28].

the pinning effect of fine Si particles on grain boundaries. Partial melting in as-cast A356 is reported to start in the temperature range of 560–563°C [66]. Therefore, the largest superplasticity observed in FSP A356 was not associated with the appearance of a liquid phase as reported for aluminum matrix composites [62–64]. Thus, the development of superplasticity in FSP A356 was attributed to the fine-grained microstructure created by FSP [21].

As presented in Chapter 3, high-strain-rate superplasticity (HSRS) has been observed in several FSP aluminum alloys such as 7075Al, Al–Mg–Zr, and 2024Al. However, FSP A356 did not exhibit HSRS, though it had fine-grained microstructure of ~3 μm (Figure 5.1(c)). This is likely that the Si particles influence the superplastic deformation process. The Si particles are ~0.3 μm and diffusional accommodation around these particles might limit the optimum superplastic strain rate.

Figure 5.3 shows tested specimens of the FSP A356 deformed to failure at 530°C and two strain rates. The specimens show neck-free elongation characteristic of superplastic flow. No noticeable deformation occurred in the grip section. This is consistent with a previous investigation on FSP Al–4Mg–1Zr.

Figure 5.4 shows the variation of flow stress with initial strain rate and temperature for both FSP and cast A356. The cast A356 exhibited a strain rate sensitivity of ~0.14 throughout the investigated strain rates of $3 \times 10^{-4} - 1 \times 10^{-1}\ s^{-1}$. This accounts for the absence of superplasticity in cast A356. By comparison, the strain rate sensitivity of the FSP sample is 0.26, 0.45, 0.43, and 0.24 in the initial strain rate ranges of

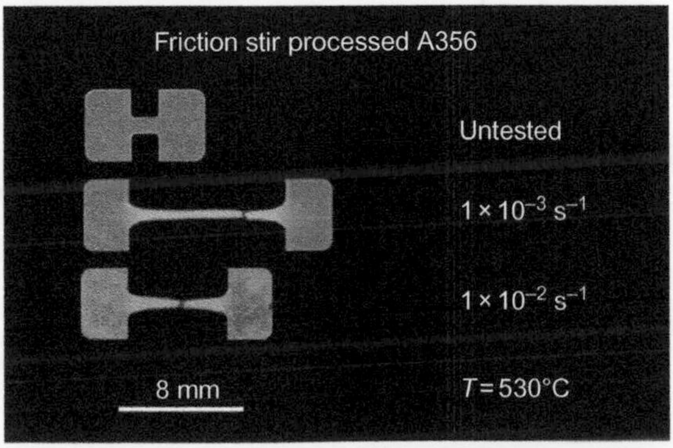

Figure 5.3 Appearance of specimens before and after superplastic deformation at 530°C [28].

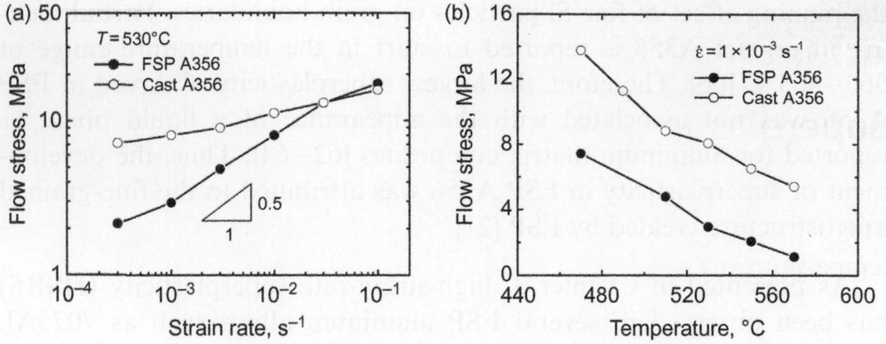

Figure 5.4 Variation of flow stress with (a) initial strain rate and (b) temperature for both FSP and cast A356 [28].

3×10^{-4}–1×10^{-3}, 1×10^{-3}–3×10^{-3}, 3×10^{-3}–3×10^{-2}, and 3×10^{-2}–1×10^{-1} s^{-1}, respectively, i.e., the FSP A356 exhibited the typical S-type stress–strain rate behavior characteristic of a superplastic material (Figure 5.4(a)). The maximum strain rate sensitivity of 0.45 at the initial strain rate of 1×10^{-3} s^{-1} corresponds to the maximum elongation of 650%. A strain rate sensitivity close to 0.5 indicates that grain boundary sliding is the dominant deformation mechanism [67]. Furthermore, flow stress of the FSP sample is significantly lower than that of the cast sample at initial strain rates $<3\times10^{-2}$ s^{-1} for 530°C and at temperatures of 470–570°C for an initial strain rate of 1×10^{-3} s^{-1}. This again is attributed to a significantly-refined microstructure in the FSP sample.

The overall implication of this study is that FSP is a very effective processing technique to create a thermally stable fine-grained microstructure in cast aluminum alloys resulting in significant superplasticity. The generation of superplasticity in A356 is likely to widen the general application of FSP for creating superplastic microstructure for selective superplasticity from cast microstructure [28].

CHAPTER 6

Superplastic Deformation Mechanism

The steady-state deformation of polycrystalline materials at elevated temperatures is usually analyzed through the equation [6,68,69]:

$$\dot{\varepsilon} = A\frac{D_0 Eb}{kT}\exp\left(-\frac{Q}{RT}\right)\left(\frac{b}{d}\right)^p\left(\frac{\sigma-\sigma_0}{E}\right)^n \tag{6.1}$$

where $\dot{\varepsilon}$ is the strain rate, A is a constant, D_0 is the preexponential factor for diffusion, E is the Young's modulus, b is the Burger's vector, k is the Boltzmann's constant, T is the absolute temperature, Q is the activation energy dependent on the rate controlling process, R is the gas constant, d is the grain size, σ is the applied stress, and σ_0 is the threshold stress.

Three variables, n, p, and Q, are the most important parameters for determining the deformation mechanism. Identifying superplastic deformation mechanism depends on the correct evaluation of the deformation parameters (n, p, and Q) and microstructural features after or during deformation, and comparison with the known physical models. In the following text, the superplastic deformation mechanism of FSP (friction stir processing) aluminum alloys will be identified by means of microstructural examinations and the evaluation of the deformation parameters.

6.1 HIGH STRAIN RATE SUPERPLASTICITY

It is widely accepted that grain boundary sliding (GBS) is the dominant superplastic deformation mechanism in fine-grained materials and is characterized by a stress exponent of ~2 [6]. As shown in Figure 3.6, for the FSP 7075Al alloys, a stress exponent of ~2 was observed in the strain rate range of 3×10^{-3}–1×10^{-1} s^{-1} at the investigated temperatures. This implies that the main mechanism for superplastic deformation in FSP 7075Al alloys is associated with GBS.

For the FSP Al–4Mg–1Zr, however, the strain rate sensitivity m increases continuously with increasing initial strain rate from 1×10^{-3}

to 1 s^{-1}, as shown in Figure 3.7. Similar behavior was also observed in a number of powder metallurgy (PM) aluminum alloys [70,71] and aluminum matrix composites [72–74]. The trend observed in the FSP Al–4Mg–1Zr indicates that either the deformation mechanism changes or a threshold stress is operative.

Al–Mg alloys are known to exhibit Class I solid behavior, namely, deformation is controlled by solute drag on gliding dislocations with a stress exponent of 3 ($m \sim 0.33$) [75–77]. For a fine-grained structure, GBS with a stress exponent of 2 ($m \sim 0.5$) is also expected to operate under certain test conditions. Since solute drag and GBS are two independent mechanisms, the deformation in Al–Mg alloys may be controlled simultaneously by both mechanisms or predominantly by one of them. A change of deformation mechanism is also expected with change in test conditions.

The solute drag mechanism in Al–Mg alloys usually takes place at intermediate temperatures around 300°C [77]. For example, in an investigation of the high temperature deformation behavior of Al–6Mg–0.3Sc alloy, Nieh et al. [78] observed that the rate controlling deformation mechanism at 350°C was solute drag controlled dislocation glide ($m \sim 0.33$), whereas the GBS process prevailed at 475°C (m close to 0.5). Because the superplastic data in fine-grained (1.5 μm) FSP Al–4Mg–1Zr were obtained at 425–580°C, solute drag is unlikely to be the rate controlling deformation mechanism.

Therefore, it is very likely that a threshold stress is operative in the FSP Al–4Mg–1Zr. The superplastic data of the FSP Al–4Mg–1Zr were examined using the threshold stress approach. Threshold stress values at various temperatures were estimated by extrapolation technique from the plots of $\dot{\varepsilon}^{1/2}$ versus the flow stress on double linear scales (Figure 6.1). The experimentally determined threshold stress decreases with increasing temperature (Figure 6.2).

SEM examinations on the surfaces of superplastically deformed specimens can provide additional information about the superplastic deformation mechanism. Figure 6.3 shows SEM micrographs of the FSP 7075Al specimens deformed at 490°C and 3×10^{-3} s^{-1} to elongations of 200%, 400%, and failure. At an elongation of 200%, SEM observations revealed evidence of GBS (Figure 6.3(a)). Some fine precipitates were faintly seen within grains. A backscattered electron image of the same region shows fine precipitates uniformly distributed within the

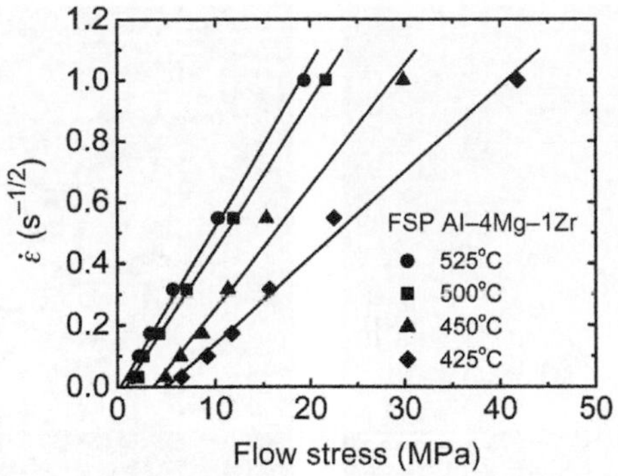

Figure 6.1 Variation of $\dot{\varepsilon}^{1/2}$ with flow stress for FSP Al–4Mg–1Zr [26].

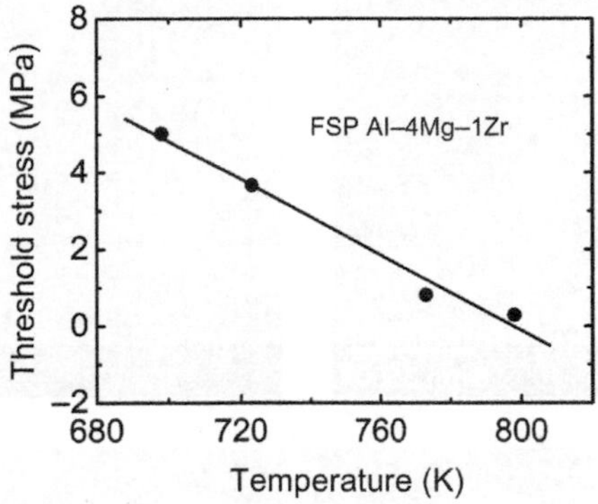

Figure 6.2 Variation of threshold stress with temperature for FSP Al–4Mg–1Zr [26].

grains (Figure 6.3(b)). Furthermore, the precipitates were distributed along the grain boundaries. Figure 6.3(c) clearly shows surface grains *A*, *B*, *C*, *D*, and *E* moving apart due to GBS. With increasing strain to an elongation of 400%, the GBS became increasingly evident and the grains were somewhat elongated along the tensile axis (Figure 6.3(d)). With increasing strain to failure, the intense GBS and grain rotation in the surface layer resulted in the appearance of a subsurface layer grain *F* on the surface and the formation of cavities (Figure 6.3(e)).

Similarly, SEM examinations on the surfaces of FSP Al–4Mg–1Zr specimens superplastically deformed at various temperatures revealed evidence of extensive GBS (Figure 6.4).

To further elucidate the superplastic deformation mechanism in the FSP aluminum alloys, superplastic data from FSP 7075Al and

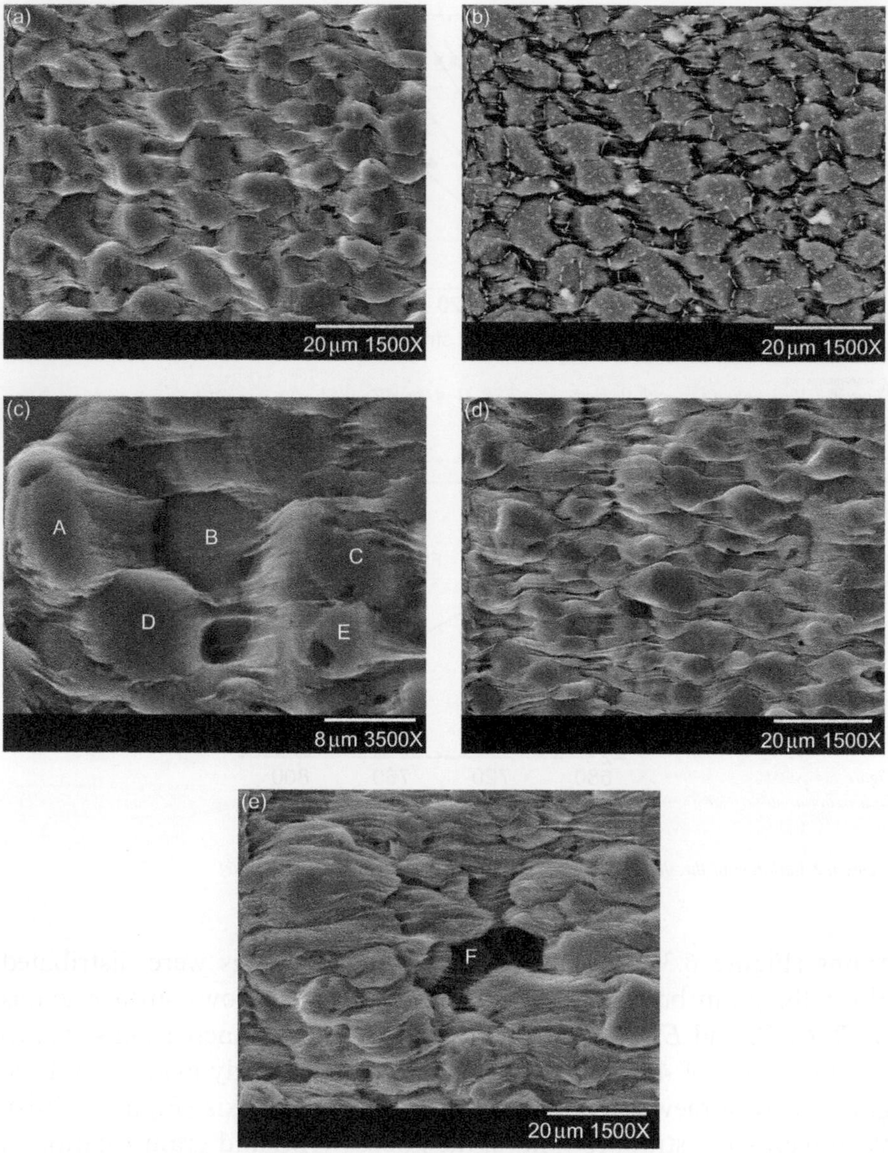

Figure 6.3 SEM micrographs showing surface morphologies of FSP 7.8 μm 7075Al superplastically deformed to (a)–(c) an elongation of 200% and (d) an elongation of 400%, and (e) failure [(a) and (b) are secondary and backscattered electron images of the same region, respectively] (tensile axis is horizontal) [24].

Al–4Mg–1Zr were plotted in Figure 6.5 as $(\dot{\varepsilon}kTd^2/(D_gEb^3))$ versus $(\sigma - \sigma_0)/E$. Three important observations can be made from this plot. First, the data of the two FSP aluminum alloys merge after the strain rates are normalized by the square of the grain sizes. This shows an

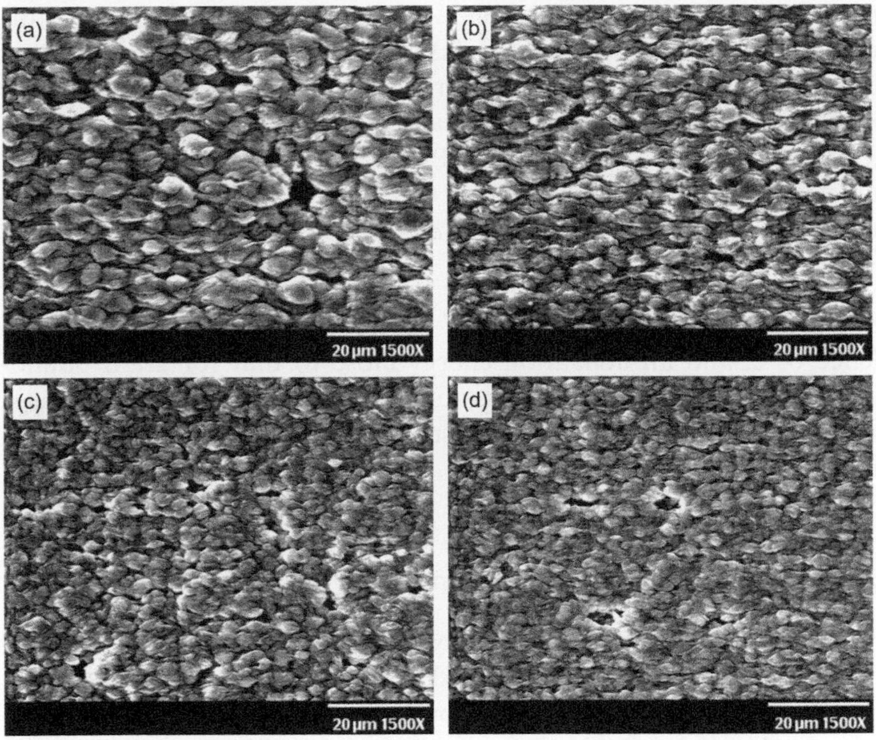

Figure 6.4 SEM micrographs showing surface morphologies of FSP Al–4Mg–1Zr superplastically deformed to failure at an initial strain rate $1 \times 10^{-1}\ s^{-1}$ and a temperature of (a) 525°C, (b) 500°C, (c) 450°C, and (d) 425°C (tensile axis is horizontal) [26].

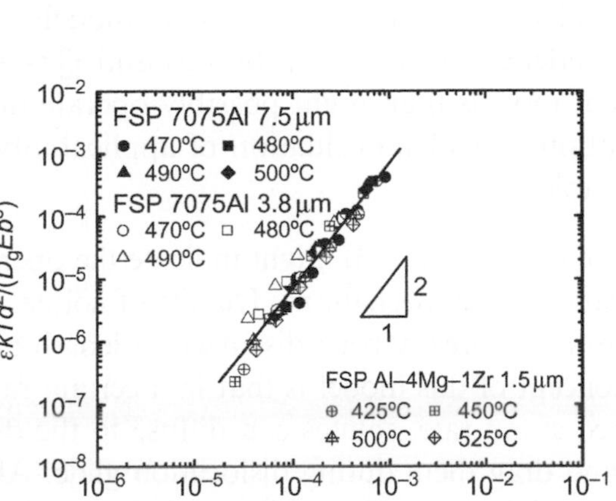

Figure 6.5 Variation of $\dot{\varepsilon}kTd^2/D_gEb^3$ with normalized effective stress, $(\sigma - \sigma_0)/E$, for FSP Al–4Mg–1Zr and 7075Al [26].

inverse grain size dependence of 2 for superplastic flow. Second, the temperature dependence of superplastic flow is similar to the activation energy for aluminum grain boundary self-diffusion. Third, irrespective of different chemistries, grain sizes and superplastic deformation characteristics, the data of the three FSP samples fits onto a single straight line with a slope of 2, showing that the stress dependence of superplastic flow is approximately 2 ($\dot{\varepsilon} \propto (\sigma - \sigma_0)^2$).

The GBS models of Mukherjee [67,79] and Ball and Hutchinson [80] predict a stress exponent of 2, an inverse grain size dependence of 2, and an activation energy close to that for grain boundary self-diffusion. Thus, Figure 6.5 shows that GBS is the main superplastic deformation mechanism for micron-structured FSP aluminum alloys.

6.2 LOW TEMPERATURE SUPERPLASTICITY

For the FSP Al–Zn–Mg–Sc alloy, an m value of $\sim$0.33 and a Q value of $\sim$142 kJ/mol were obtained in the wide range of temperatures and strain rates investigated as shown in Figure 4.3. However, the p value could not be obtained since only one grain size was evaluated in the FSP Al–Zn–Mg–Sc alloy. Hence, the approach toward elucidating deformation mechanism in the FSP Al–Zn–Mg–Sc alloy is to examine suitability of different physical models available with parametric dependencies of $m = 0.33$ ($n = 3$) and $Q = \sim Q_L$. Table 6.1 presents different models that might possibly operate during the superplastic deformation. Detailed evaluation of applicability of each model is given in the table.

An m value of 0.33 (i.e., $n = 3$) might indicate the operation of the solute drag dislocation glide mechanism. Details of solute drag dislocation glide mechanism have already been discussed at length in various papers [81–83]. The concept of this model is that in a certain range of temperatures and applied strain rate, solutes can diffuse to the dislocation cores, saturate them and drag them during dislocation glide. Alloying elements with large atomic misfit parameter (e) have a better tendency toward generating a saturated solute atmosphere around the dislocations, and thus, the glide becomes the rate limiting step. However, operation of solute drag creep has been found to exhibit at most 400% elongation, in certain Al–Mg alloys. In those cases, almost all of the strain comes from the elongation of the individual deforming grains. For example, Pu et al. [5] and Hsiao and Huang [11] described LTSP (low temperature superplasticity) mechanism in terms of solute drag dislocation glide, yet the ductility values in their studies were moderate. It should be pointed out that superplastic elongation as high as 1165% was achieved in the FSP Al–Zn–Mg–Sc alloy. Furthermore, SEM observations revealed evidence of near-equiaxed grains on the surfaces of superplastically deformed specimens (Figure 6.6). Thus, significant superplastic elongation

Table 6.1 A Summary for Superplastic Mechanisms Relating Deformation Mechanism in the FSP Alloy

Physical Mechanism	Predicted Parametric Dependencies	Predicted Microstructural Features	Experimental Observations	Comments
Solute drag dislocation creep [131]	$n = 3, p = 0, D = D_{chem}$	Elongated grains, moderate elongations, dislocation network formation (no subgrain formation)		Not possible
Viscous glide limited GBS [84]	$n = 1, p = 2, D = D_{chem}$	Equiaxed grains, minimal dislocation activity	Almost equiaxed morphology, grains	Not possible
Viscous glide (particle shearing) limited GBS [85]	$n = 2, p = 1, D = D_{gb}$	Equiaxed grains, minimal dislocation activity	Sliding against each other $n = 3, D = D_L$	Not possible
Interface reaction controlled GBS [89]	$n = 3, p = 1, D = D_{chem}$	Equiaxed grains, minimal dislocation activity, true GBS appears at higher strain rates		Not possible
Rachinger sliding ($d > \lambda$) [89]	$n = 3, p = 1, D = D_L$	No grain elongation (equiaxed morphology)		Possible

Figure 6.6 Surface topography of a superplastically deformed tensile specimen (310°C, 3×10^{-2} s^{-1}) near the fracture tip (tensile axis is horizontal) [31].

and the topographical microstructure of near-equiaxed grains preclude the suitability of the solute drag model in its classical form for the FSP alloy.

In view of the large elongation and the topography of near-equiaxed grains in the FSP Al–Zn–Mg–Sc alloy, some GBS-related deformation mechanisms where grains are allowed to slide against each other must be invoked to explain the phenomenon. A GBS model proposed by Fukuyo et al. [84] involving dislocation glide limited GBS concept is included. The drawback of this model is the prediction of linear stress dependence of strain rate ($n = 1$) and $p = 2$. Also, a derivative of this model where cutting of coherent particles in the matrix occurs during viscous glide is unacceptable because it predicts $n = 2$ with $p = 1$ [85]. Hence, this model could not be applicable for describing the superplastic behavior of the FSP Al–Zn–Mg–Sc alloy as their parametric dependencies are at odds with the experimental observations. Most physical models for superplasticity involving GBS predict an m value of 0.5 (i.e., $n = 2$), Q close to either Q_L (lattice) or Q_{GB} (grain boundary) and $p = 2$ or 3 [86]. It is important to note that no known model predicts $n = 3$, $p = 2$ or 3 (irrespective of which Q is applicable). However, a parametric combination of $n = 3$, $p = 1$, and $Q = Q_L$ offers a more plausible scenario. Hence, two possible GBS-related mechanisms with similar parametric dependencies are considered.

The first one is impurity controlled or interface reaction controlled GBS. Experimental evidence for superplastic yttria-stabilized tetragonal zirconia (3YTZ) revealed that there might be an impurity controlled or interface reaction controlled GBS regime, which acts sequentially with GBS at higher stresses [87,88]. Langdon [89] noted that this regime might have a stress exponent value of 3 (i.e., characteristic of viscous glide), p of 1 considering the major strain contributed by GBS, and $D = D_{sol}$ where D_{sol} is the appropriate diffusivity for the solute causing the drag effect on the dislocations. However, the exact nature of this model remains unclear. Also, it could be noted that it is experimentally validated only in 3YTZ (without amorphous grain boundary phase) ceramics, not in metallic alloys. Considering the strain rate and temperature conditions, the chances of such GBS regime appearing in the FSP Al–Zn–Mg–Sc alloy are low, since at higher strain rates thermally activated glide in place of diffusion-controlled mechanisms would probably start to operate.

The second possible GBS-related mechanism is Rachinger GBS. Rachinger [90] first described that strain can be accumulated in a specimen through grain rearrangement during GBS in such a way that the number of grains in the cross-sectional area changes. Later, Langdon [91] developed a model based upon the Ball–Hutchison superplasticity model for Rachinger sliding accommodated by intragranular slip. It was shown that for $d > \lambda$ (subgrain size), the constitutive equation for Rachinger sliding is given by [91]

$$\varepsilon^{\bullet}_{\text{gbs}} = \frac{A_{\text{gbs}(d>\lambda)} D_{\text{L}} E b}{kT} \left(\frac{\sigma}{E}\right)^{3} \left(\frac{b}{d}\right)^{1} \tag{6.2}$$

where $A_{\text{gbs}(d>\lambda)}$ is a dimensionless constant and is predicted to be ~7100 [91]. Superplastic data of the FSP Al–Zn–Mg–Sc alloy are plotted in a normalized manner in Figure 6.7 and compared with Eq. (6.2). It shows that the constant "A" for the FSP alloy is $\sim 5.1 \times 10^6$, which keeps the data more than an order of magnitude apart on the normalized strain rate scale from the theoretical equation. At the same time, it can be noted that the constant value was predicted for the Rachinger sliding occurring for commercially pure Al under creep conditions. Therefore, further information on texture and microstructural analysis of deformed tensile samples might determine whether this mechanism would best explain the experimental findings in the FSP Al–Zn–Mg–Sc alloy.

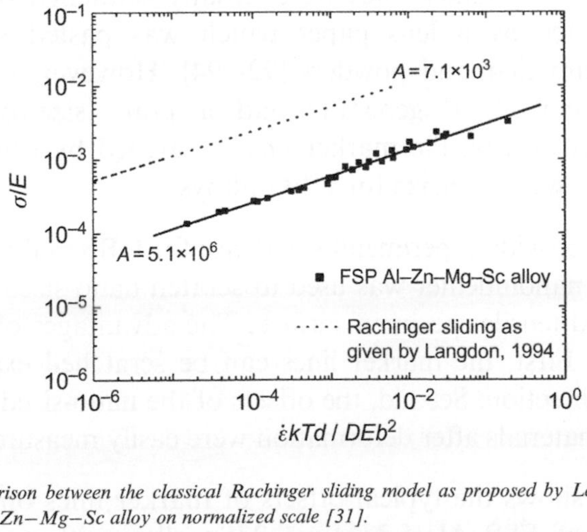

Figure 6.7 A comparison between the classical Rachinger sliding model as proposed by Langdon [91] and the data of the FSP Al–Zn–Mg–Sc alloy on normalized scale [31].

Similarly, for ultrafine-grained (UFG) FSP Al–4Mg–1Zr alloy, an *m* value of 0.3–0.4 was observed for the optimum superplasticity, as shown in Figure 4.7(b). This implies that main deformation mechanism for LTSP of UFG aluminum alloys might be associated with solute drag considering the fact that the aluminum alloys exhibiting LTSP are Mg-containing alloys. From Table 4.1, it can be seen that the *m* values for LTSP of UFG aluminum alloys prepared by various techniques are generally between 0.3 and 0.4, lower than the characteristic sensitivity for GBS, which is 0.5. The LTSP deformation mechanism of UFG aluminum alloys is still not well understood in spite of a great number of reports about LTSP of aluminum alloys. Pu et al. [5] suggested that viscous dislocation creep would control the LTSP flow of 8090Al alloy. However, Park et al. [13] claimed that the LTSP of the ECAP UFG 5083Al was attributed to GBS, which was rate controlled by grain boundary diffusion. Hsiao et al. [11] observed that GBS did occur in TMT (thermo-mechanically treated) 5083Al at temperatures as low as 200°C. During the initial LTSP stage, the primary deformation mechanisms were solute drag creep plus minor power law creep. At later stages, GBS gradually controlled the deformation [11].

Clearly, there is still no consensus on the deformation mechanism of LTSP. The main reason for this situation is a lack of experimental evidence. Measuring the offsets of marker lines on the surfaces of deformed specimens is a direct and effective approach to estimating the contribution of GBS to the total strain [92–94]. For micron-grained alloys, the maker lines were usually scratched on the tensile specimen surface by a lens paper which was pasted with a small amount of 3 μm diamond powders [92–94]. However, the aluminum alloys exhibiting LTSP generally had a grain size of submicron (Table 4.1). Obviously, the marker lines scratched by using 3 μm diamond powders were oversize for UFG alloys.

In order to provide experimental evidence for GBS in the UFG aluminum alloys, a nanoindenter was used to scratch nano-sized marker lines on the polished tensile specimen surface. The advantages of this method were obvious. First, the marker lines can be scratched exactly parallel to the tensile direction. Second, the offsets of the nano-sized marker lines on the UFG materials after deformation were easily measured.

Figure 6.8 shows the typical offsets of marker lines on the surfaces of specimens of FSP Al–5.33Mg–0.23Sc alloy with a grain size of ~0.6 μm deformed at 175°C and $1 \times 10^{-4}\ s^{-1}$ to different elongations.

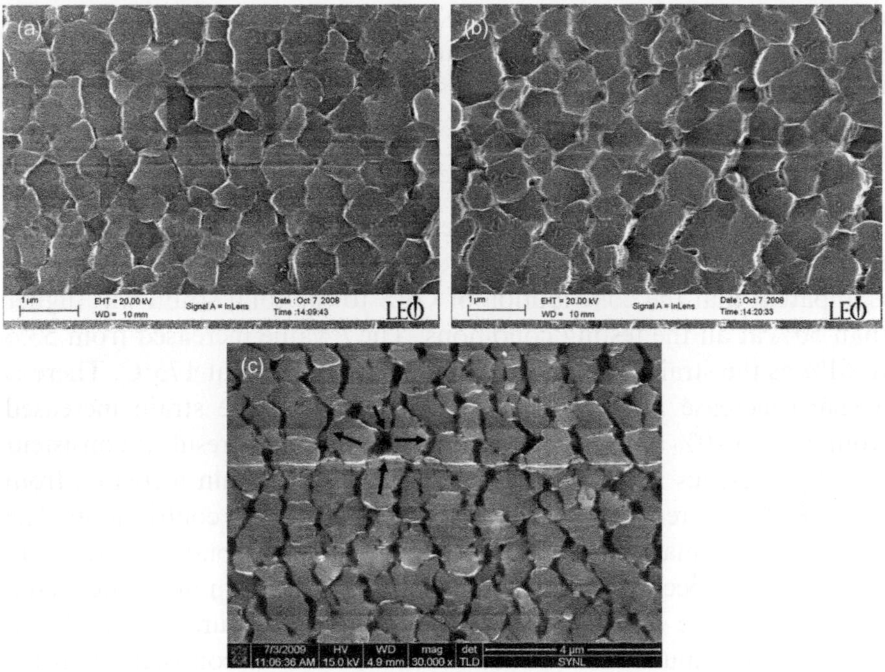

Figure 6.8 SEM micrographs showing the offsets of marker lines in 0.6 μm FSP Al–Mg–Sc alloy specimens deformed at 175°C and $1 \times 10^{-4}\ s^{-1}$ to elongations of (a) ~20%, (b) ~40%, and (c) ~80% (the loading direction is horizontal) [141].

The grains remained equiaxed in shape during deformation. The marker lines were sharply defined and exhibited distinct sliding offsets at many grain boundaries. The sliding offsets tended to increase with an increase in elongation. When being pulled to an elongation of 80%, the specimen showed the evidence of grain rotation and the development of cavities at the grain triple junctions or the grain boundaries. Most of the cavities tended to develop perpendicular to the tensile direction. Further, the sliding directions of some grains were clearly visible as indicated by the arrows in Figure 6.8(c).

The GBS contributions were determined by measuring the sliding offsets (w) combined with measurements of the mean linear intercept grain sizes (L) [95].

$$\varepsilon_{\text{gbs}} = \phi \frac{\overline{w}}{\overline{L}} \tag{6.3}$$

where $\overline{w}$ is the average value of w, $\overline{L}$ is the mean linear intercept grain size, and ϕ is a constant with a value which was estimated both experimentally and theoretically, to be ~1.5. If ε_{total} denotes the total strain in the specimen, the contribution of GBS to the strain, ξ, may be expressed as a fractional relationship [96].

$$\xi = \frac{\varepsilon_{gbs}}{\varepsilon_{total}} \tag{6.4}$$

Figure 6.9 shows the measurement results of the GBS contribution to the strain at different temperatures after tested to different strains. It is apparent that the contribution of GBS to the total strain was higher than 50% at all the testing conditions. The ξ value increased from 53% to 61% as the strain was increased from 20% to 80% at 175°C. There is a sharp increase in the GBS contribution when the strain increased from 20% to 40% at the test temperature range. This result is consistent with the previous report [11]. However, when the strain increased from 40% to 80%, there is only a slight increase in the GBS contribution. The reason for this may be that besides GBS, the grain rotation and cavity formation also occurred at higher strain range as shown in Figure 6.8(c). Furthermore, the ξ values increased as the temperature increased and reached a maximum value of 72% at 300°C. This is consistent with the result that higher elongation was obtained at higher temperature.

Clearly, GBS did operate at temperatures as low as 175°C in the UFG FSP Al–Mg–Sc, especially for the later straining stage, though the strain rate sensitivity is only ~0.33. The occurrence of GBS during high temperature plastic flow is closely related to the grain boundary

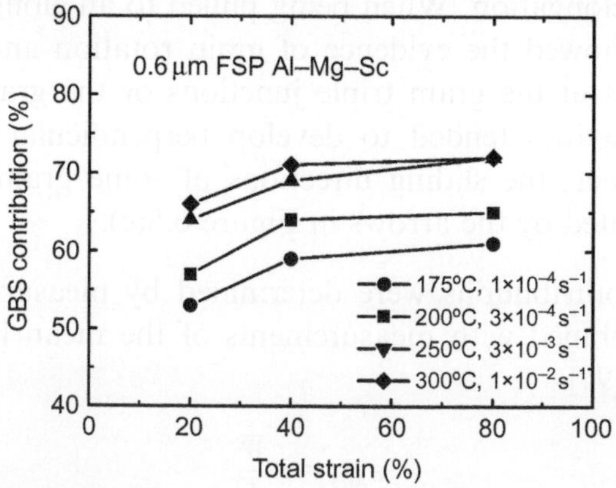

Figure 6.9 The GBS contribution to the total strain of 0.6 μm FSP Al–Mg–Sc alloy at different temperatures after tested to different strains [141].

characters [97,98]. It is generally believed that GBS occurs along the HAGBs (high-angle grain boundaries), whereas LAGBs (low-angle grain boundaries) are considered immobile with respect to grain sliding [99]. The fraction of the HAGBs as high as 95% in the UFG FSP Al–Mg–Sc is significantly higher than that in conventional TMT and ECAP (equal channel angular pressing) aluminum alloys with a typical HAGB ratio of 50–65% [9–18]. Exceptionally high ratio of the HAGBs made GBS take place easily and the contribution of GBS to the total strain increase, resulting in the occurrence of superplasticity at temperature as low as 175°C in the UFG FSP Al–Mg–Sc with GBS as a primary superplastic deformation mechanism.

As shown in Figure 4.9, for the UFG FSP UFG Al–4Mg–1Zr, the *m* values tended to increase with increasing the strain rate in regions I and II over temperatures ranging from 175 to 425°C, like the trend observed in the FSP fine-grained (1.5 μm) Al–4Mg–1Zr (Figure 3.7). This trend in the variation of the *m* value indicated that either a threshold stress was operative or the deformation mechanism had changed. It has been suggested that lower *m* values in region I did not represent a genuine change in the rate controlling mechanism, but rather originated from the existence of a threshold stress due to segregation of impurity atoms on the grain boundaries [100,101]. If a threshold stress is responsible for the change in the *m* value, it can be determined when the deformation mechanism is known or assumed. As shown in Figures 6.5 and 6.6, GBS was the predominant deformation mechanism for FSP UFG Al–5.33Mg–0.23Sc even at 175°C. Therefore, it is reasonable to use a true stress exponent of 2 ($m = 0.5$) to analyze the superplastic data of UFG FSP UFG Al–4Mg–1Zr.

In Figure 6.10, plots of σ against $\dot{\varepsilon}^{1/2}$ ($n = 2$) are presented on double linear scales by using the superplastic data in regions I and II for all testing temperatures. By an extrapolation of the data to zero strain rate with a linear regression, all values of threshold stresses could be estimated. The calculated threshold stresses are summarized in Table 6.2. The threshold stress values were highly dependent on the deformation temperature. Furthermore, the observed linear behavior on the plot of σ against $\dot{\varepsilon}^{1/2}$ indicates that the true *m* value is 0.5.

The activation energy, which depends on a rate controlling process, is also a very important factor in the determination of the

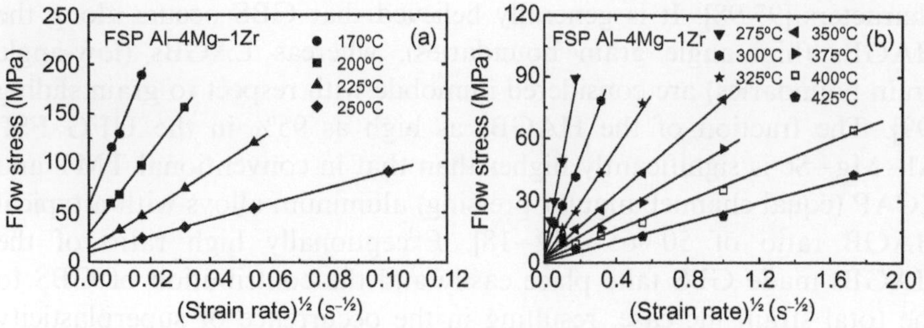

Figure 6.10 Variation of flow stress as a function of $\dot{\varepsilon}^{1/2}$ for FSP Al–4Mg–1Zr [142].

Table 6.2 Threshold Stress of FSP Al–4Mg–1Zr Sample at Various Temperatures [142]

Temperature (°C)	175	200	225	250	275	300	325	350	375	400	425
Threshold stress (MPa)	49.31	27.67	13.00	10.16	4.82	4.19	2.50	2.46	2.17	1.62	0.68

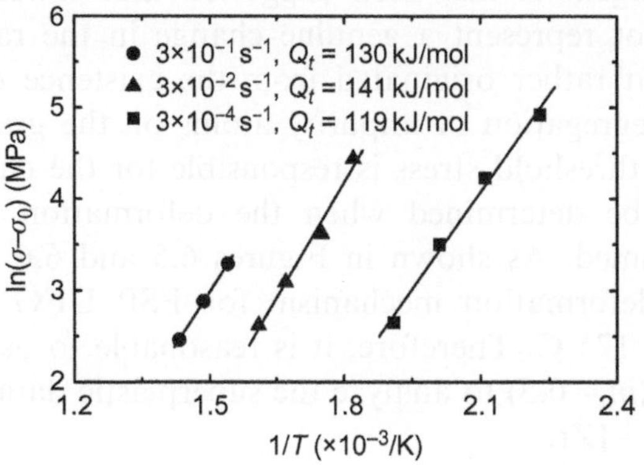

Figure 6.11 Variation of $\ln(\sigma - \sigma_0)$ as a function of reciprocal temperature for FSP Al–4Mg–1Zr [142].

deformation mechanism. The true activation energies evaluated according to the following equation [11] are shown in Figure 6.11.

$$Q_t = nR\frac{\partial[\ln(\sigma - \sigma_{\text{th}})]}{\partial(1/T)}\Big|_{\dot{\varepsilon}} \tag{6.5}$$

The estimated activation energies under the constant strain rates are ~119–141 kJ/mol. These values of Q_t are much higher than those for dislocation pipe diffusion in Al (82 kJ/mol) and grain boundary diffusion in Al (84 kJ/mol), and close to those for the Mg diffusion in the Al matrix (136 kJ/mol) and lattice self-diffusion of pure Al (142 kJ/mol). Therefore, the most acceptable rate controlling diffusion step might be the lattice diffusion of pure Al. This is similar to that observed in the FSP UFG Al–Zn–Mg–Sc alloy.

At a constant temperature (T) and a normalized stress ($\sigma - \sigma_0$), for discussion of the relationship between grain size and flow stress, Eq. (4.1) can be rewritten as

$$\ln \dot{\varepsilon} = A' + p \ln\left(\frac{b}{d}\right) + n \ln\left(\frac{\sigma - \sigma_0}{E}\right) \tag{6.6}$$

where A' is a temperature-dependent coefficient. The p value can be determined by finding the relationship between $\dot{\varepsilon}$ and d at a constant ($\sigma - \sigma_0$) value. The superplastic data of the micron-grained (1.5 μm) Al–4Mg–1Zr (Figure 3.7) were used in this study to determine the p value. The double linear plot of $\ln \dot{\varepsilon}$ against $\ln(b/d)$ is presented in Figure 6.12. The p value was determined to be close to 2.

To further elucidate the superplastic deformation mechanism in the FSP UFG Al–4Mg–1Zr, superplastic data are plotted in Figure 6.13 as $\dot{\varepsilon}kTd^2/(G_L E b^3)$ versus $(\sigma - \sigma_0)/E$. It can be seen that all data for the FSP UFG Al–4Mg–1Zr fit onto a single straight line with a slope of 2 after introducing the threshold stress, showing that a threshold type deformation behavior with a stress exponent of 2 ($\dot{\varepsilon} \propto (\sigma - \sigma_0)^2$). The activation energy value, which is dependent on the rate controlling

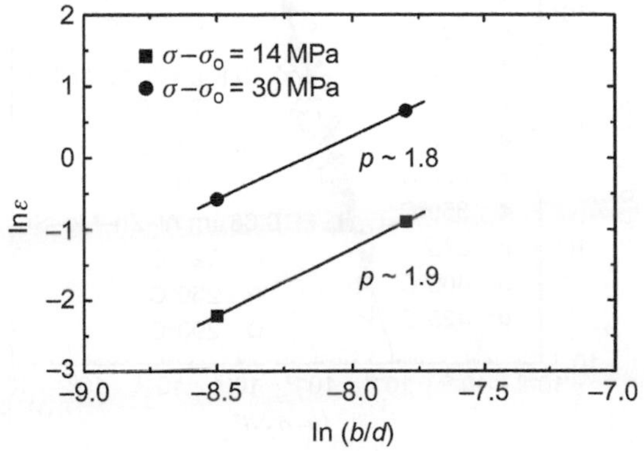

Figure 6.12 Variation of ln $\dot{\varepsilon}$ as a function of ln(b/d) for FSP Al–4Mg–1Zr [142].

process, is similar to the lattice self-diffusion of pure Al (142 kJ/mol). The constitutive equation in the FSP UFG Al–4Mg–1Zr is given as

$$\dot{\varepsilon} = 5 \times 10^7 \frac{D_0 E b}{kT} \exp\left(-\frac{142{,}000}{RT}\right)\left(\frac{b}{d}\right)^2\left(\frac{\sigma-\sigma_0}{E}\right)^2 \tag{6.7}$$

This indicates that the dominant deformation mechanism for the FSP UFG Al–4Mg–1Zr is GBS controlled by the lattice diffusion.

For comparison, the data for the FSP UFG aluminum alloys from other studies [31,54,55] are also included in Figure 6.13. At high temperatures, the actual grain sizes before superplastic deformation were much larger than the initial ones for these UFG alloys. Therefore, only the superplastic data obtained at $\leq 300°C$ are added in this plot. It is clear that the present Al–4Mg–1Zr alloy behaved identically to other UFG superplastic aluminum alloys.

6.3 ENHANCED DEFORMATION KINETICS

The GBS models of Mukherjee [67], Arieli and Mukherjee [79], and Ball and Hutchison [80] predict a stress exponent of 2, an inverse grain size dependence of 2, and an activation energy close to that for grain boundary self-diffusion with the dimensionless constant *A* being within the range of 25–50.

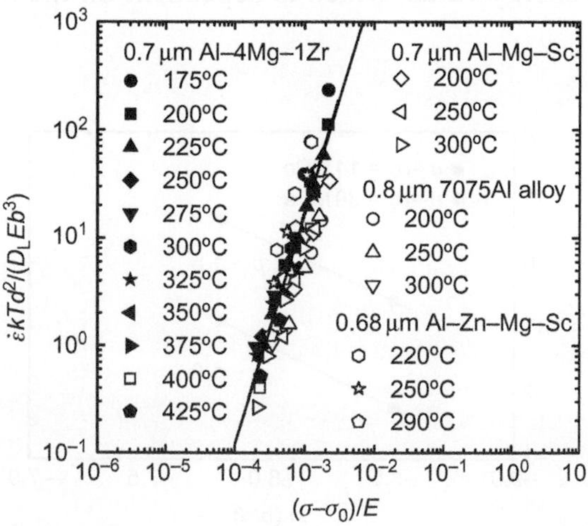

Figure 6.13 Variation of $(\dot{\varepsilon}kTd^2/D_g E b^3)$ as a function of $(\sigma - \sigma_0)$/E for FSP UFG aluminum alloys [142].

Mishra et al. [6] analyzed superplasticity data for a number of PM aluminum alloys. They observed a threshold stress type behavior in aluminum alloys with second phase particles. In this case, the superplastic deformation behavior of these aluminum alloys can be described by a modified relationship [6]:

$$\dot{\varepsilon} = 40\frac{D_0 Eb}{kT}\exp\left(-\frac{84,000}{RT}\right)\left(\frac{b}{d}\right)^2\left(\frac{\sigma-\sigma_0}{E}\right)^2 \tag{6.8}$$

where D_0 is the preexponential constant for grain boundary diffusivity, E is the Young's modulus, R is the gas constant, and σ_0 is the threshold stress. By comparing the GBS model of Mishra et al. [6] with that of Mukherjee [67], Arieli and Mukherjee [79], and Ball and Hutchison [80], it is clear that both models are consistent with similar A value.

In Figure 6.14, the superplastic data of FSP 7075Al and Al–4Mg–1Zr were compared with a prediction by Eq. (6.8) as $(\dot{\varepsilon}kTd^2/(D_gEb^3))$ versus $(\sigma - \sigma_0)/E$. The superplastic data of the two alloys can be described by

$$\dot{\varepsilon} = 700\frac{D_0 Eb}{kT}\exp\left(-\frac{84,000}{RT}\right)\left(\frac{b}{d}\right)^2\left(\frac{\sigma-\sigma_0}{E}\right)^2 \tag{6.9}$$

However, the dimensionless constant in Eq. (6.9) is more than 1 order of magnitude larger than that in GBS models of Mukherjee [67], Arieli and Mukherjee [79], and Mishra et al. [6]. This indicates that the

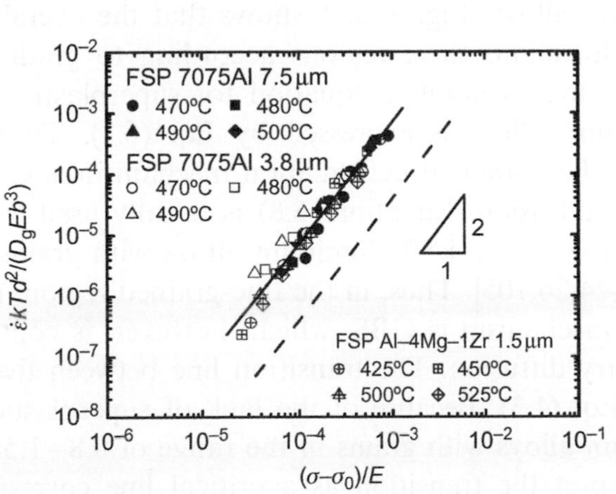

Figure 6.14 Variation of $\dot{\varepsilon}kTd^2/D_gEb^3$ with normalized effective stress, $(\sigma - \sigma_0)$/E for FSP Al–4Mg–1Zr and 7075Al (dashed line represents Eq. (6.1)) [26].

FSP aluminum alloys exhibited enhanced kinetics compared to the prediction by Eq. (6.8). Clearly, conventional GBS models are not able to predict the deformation kinetics in FSP aluminum alloys although the parametric dependencies are consistent.

As pointed out previously, the microstructure of FSP aluminum alloys is characterized by fine and high-angled grains (Figures 2.2 and 4.5). The fraction of high-angle boundaries is as high as 85–95% [23,29,102]. This ratio is significantly higher than that obtained in conventional TMP (thermo-mechanically processed) aluminum alloys with a typical ratio of 50–65% [103,104]. A high percentage of high-angle boundaries makes the contribution of GBS to strain significantly higher, thereby resulting in enhanced kinetics. It is proposed that the dimensionless constant in Eq. (4.1) is a function of the fraction of high-angle boundaries and increases as the fraction of high-angle boundaries increases. Attempts are required to correlate the dimensionless constant in Eq. (4.1) with the fraction of high-angle boundaries to establish a better constitutive relationship for superplastic deformation of fine-grained aluminum alloys.

6.4 SUPERPLASTIC MECHANISM MAP FOR FSP ALUMINUM ALLOYS

Eqs. (4.3), (6.7), and (6.8) and the reported superplastic data have made it possible to construct a superplastic mechanism map for the FSP aluminum alloys. Figure 6.15 shows that the overall superplastic region is divided into three regions according to grain size. As discussed above, the constitutive equation for superplasticity in the FSP UFG aluminum alloys is expressed by Eq. (6.7). Therefore, in the UFG region, the dominant deformation mechanism is GBS controlled by the lattice diffusion. Equation (6.8) is usually used to describe the superplastic flow of the FSP aluminum alloys with grains in the range of 1–10 μm [24,26,105]. Thus, in the fine-grained region, the dominant deformation mechanism is GBS, which, however, is controlled by the grain boundary diffusion. The transition line between the two regions depends on Eq. (4.3). Because of the lack of superplastic data of the FSP aluminum alloys with grains in the range of 0.8–1.5 μm, it is still hard to interpret the transition as a critical line corresponding to a well-defined grain size or as a region corresponding to a grain size range. Previous studies have indicated that the principal mechanism of superplasticity in coarse-grained Al–Mg alloys is solute drag on gliding dislocations [106]. For lack of superplastic data of the coarse-grained FSP aluminum alloys, the plastic deformation mechanism for the coarse-grained FSP aluminum alloys at high temperatures is unknown. AGG region is also shown in Figure 6.15 based on the reported results. Reducing grain size usually decreased AGG to lower temperature. Figure 6.15 also indicates that grain refinement is beneficial to achieving superplasticity at higher strain rate and/or lower temperature.

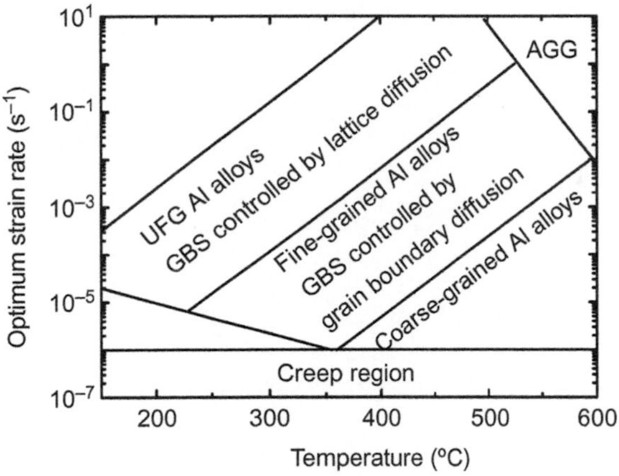

Figure 6.15 A superplastic mechanism map for FSP aluminum alloys [142].

CHAPTER 7

Cavitation During Superplasticity

It is well established that cavitation can occur in a wide range of materials during tensile superplastic flow. In recent years, extensive attention has been paid to the cavitation behavior of superplastic alloys because cavitation leads to degradation of the overall properties of the post-SPF materials [107,108]. It has been demonstrated that the post-SPF mechanical properties of the materials are significantly reduced when the cavity volume fraction exceeds approximately 1% [109].

According to Stowell et al. [110], the critical strain rate, $\dot{\varepsilon}_c$, below which cavity nucleation at a grain boundary particle of radius, r_p, is likely to be inhibited by diffusive stress relaxation, is expressed as

$$\dot{\varepsilon}_c = \frac{2.9\sigma\Omega\delta D_b}{\alpha d r_p^2 kT} \tag{7.1}$$

where σ is the stress required to nucleate a stable cavity, Ω is the atomic volume, δD_b is the product of grain boundary width and diffusivity, α is the fraction of the total strain attributable to the grain boundary sliding (GBS), d is the grain diameter, k is Boltzmann's constant, and T is the absolute temperature.

Equation (7.1) predicts that cavitation will be minimized when both grain and particle sizes are small. Therefore, development of fine-grained microstructure in commercial aluminum alloys is highly desirable for both enhanced superplasticity and reduced cavity level. For this purpose, various processing techniques such as thermo-mechanical processing (TMP) [1,3], equal channel angular pressing (ECAP) [8,111], and torsional deformation under pressure [112] were developed to produce fine-grained aluminum alloys. As discussed in Chapter 1, fine-grained structure with fine secondary phase particles was produced in the aluminum alloys via friction stir processing (FSP) (Figures 2.1 and 2.3). It is expected that such fine-grained structure will exhibit lower cavity level during superplastic deformation.

7.1 CAVITY FORMATION AND GROWTH

Figure 7.1 shows the effect of true strain on the cavitation of FSP 7.5 μm 7075Al alloy deformed at 480°C and an initial strain rate of $1 \times 10^{-2}\ s^{-1}$. At a true strain of 1.0, the development of cavities was distinctly detected (Figure 7.1(a)). The increase in the true strain resulted in the growth of the cavities and the formation of more new cavities, leading to the increase in the density and volume fraction of cavities (Figure 7.1(b) and (c)). At the true strain of 1.8, it seems that the cavity density did not increase further (Figure 7.1(d)). However, the growth, coalescence, and linkage of cavities became increasingly evident, resulting in an increase in the cavity volume fraction. It is noted that the cavities generally exhibited an irregular shape, indicating that the nucleation and growth of cavities were associated with the grain triple junctions and grain boundaries. SEM examinations revealed similar phenomenon (Figure 7.2). In addition, the formation of cavities was also associated with the coarse particles as evidenced by SEM observations.

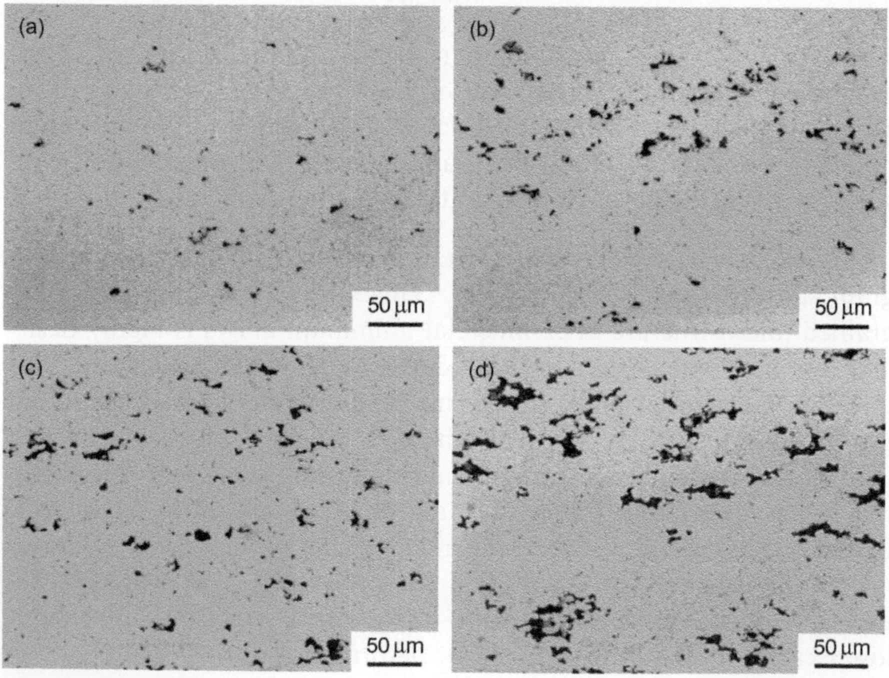

Figure 7.1 Optical micrographs showing cavitation of FSP 7.5 μm 7075Al alloy deformed at 480°C and an initial strain rate of $1 \times 10^{-2}\ s^{-1}$ to a true strain of (a) 1.0, (b) 1.3, (c) 1.5, and (d) 1.8 (tensile axis is horizontal) [140].

Figure 7.3 shows the variation of cavity density with true strain for two FSP 7075Al alloys deformed at 480°C and an initial strain rate of $1 \times 10^{-2}\ s^{-1}$. The cavity density increased rapidly with increasing strain in the intermediate stage of superplastic deformation for both FSP 7075Al alloys. This indicates that new cavities were continuously nucleated in the intermediate stage of deformation with increasing strain. In this stage, both cavity growth and nucleation of new cavities contributed to increasing cavity volume fraction. However, when strain reached a certain value, the cavity density tended to decrease for both alloys. This showed that the effect of coalescence and linkage of cavities exceeded that of nucleation of new cavities at large strain. In this

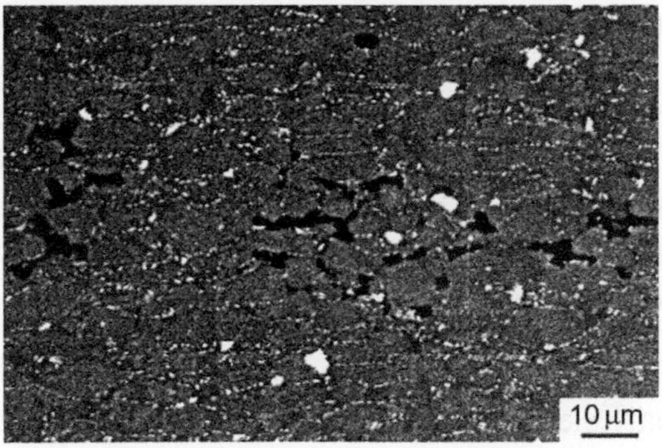

Figure 7.2 SEM micrographs showing cavity formation at large particles and grain triple junctions in FSP 3.5 μm 7075Al alloy deformed at 480°C and an initial strain rate of $1 \times 10^{-2}\ s^{-1}$ to a true strain of ~1.8 (tensile axis is horizontal) [140].

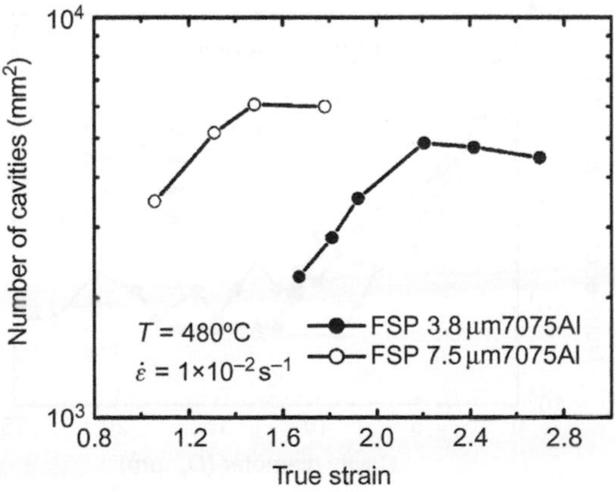

Figure 7.3 Variation of cavity density with true strain for FSP 7.5 and 3.8 μm 7075Al alloys deformed at 480°C and an initial strain rate of $1 \times 10^{-2}\ s^{-1}$ [140].

case, the increase in the cavity volume fraction mainly resulted from the growth, coalescence, and linkage of cavities. Two-dimensional cavity size distribution indicated substantial increase in number of the large size cavities with increasing strain (Figure 7.4). This means that while new cavities were continuously nucleated during superplastic deformation, the increase in the strain resulted in significant growth, coalescence, and linkage of cavities.

It is well accepted that the GBS is the dominant mechanism of superplastic deformation in the superplastic materials, and when the GBS cannot be well accommodated by other mechanisms, e.g., grain boundary migration, diffusion flow, or dislocation slip, stress concentration at certain sites may cause the development of cavitation. The triple junctions of grains, particles, and even ledges on grain boundary can serve as the nucleate sites of cavities [113,114]. Bae and Ghosh [115] reported that most cavities in a superplastic Al–Mg alloy were observed at the interfaces between particles and aluminum matrix, indicating that the cavities were formed at the particle–matrix interfaces located at grain boundaries.

For the FSP 7075Al alloys, although cavities were found to initiate at the large particles (Figure 7.2), the cavitation was more often observed to develop at the grain triple junctions as evidenced clearly by SEM examinations (Figure 7.2) and by the jagged cavity shapes (Figure 7.1). This is attributed to the fine and relatively uniformly distributed precipitates in the FSP alloys. The cavity nucleation at grain

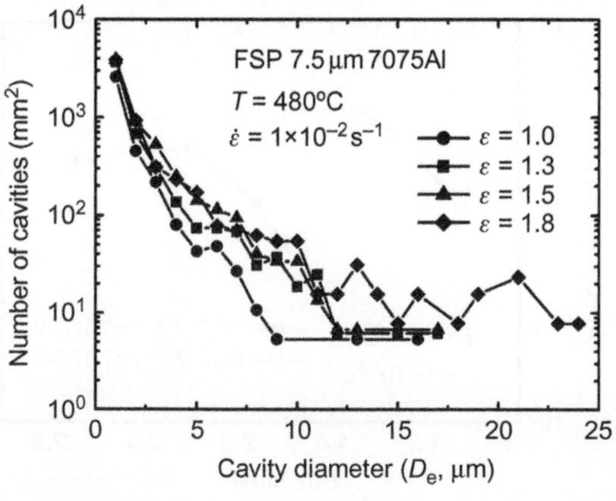

Figure 7.4 Variation of cavity density with cavity diameter for FSP 7.5 μm 7075Al alloy deformed at 480°C and an initial strain rate of $1 \times 10^{-2}\ s^{-1}$ to different strains [140].

boundary particles was deferred, resulting in an increased possibility of cavity nucleation at the grain triple junctions. Thus, the FSP 7075Al alloys could endure more strain before the cavities were extensively nucleated.

7.2 FACTOR INFLUENCING CAVITY FORMATION AND GROWTH

7.2.1 Strain Rate

Figure 7.5 shows the effect of initial strain rate on the cavitation of FSP 3.8 μm 7075Al alloy deformed at 480°C to a true strain of 1.5. At $1\times10^{-2}\,s^{-1}$, except for few small preexisting cavities, there was no distinct evidence of cavity formation (Figure 7.5(a)). With increasing initial strain rate to $3\times10^{-2}\,s^{-1}$, development of a few cavities was observed (Figure 7.5(b)), whereas at a high initial strain rate of $1\times10^{-1}\,s^{-1}$, a large number of cavities were generated in the specimen (Figure 7.5(c)). Clearly, the number of cavities increased with increasing initial strain rates at a given strain. Furthermore, it is evident from Figure 7.5 that the cavity size increases considerably with

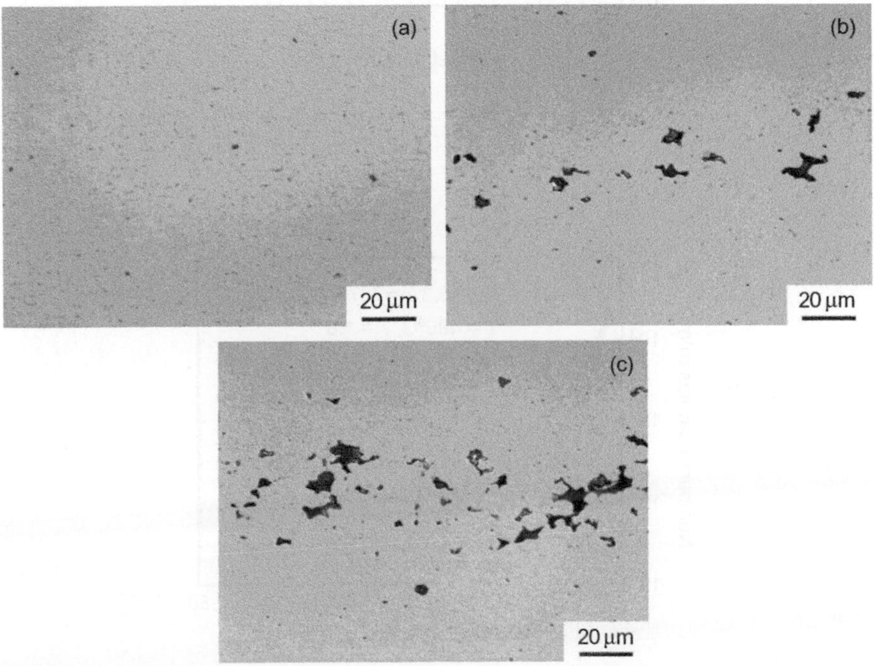

Figure 7.5 Optical micrographs showing cavitation of FSP 3.5 μm 7075Al alloy deformed at 480°C to a true strain of ~1.5: (a) $\dot{\varepsilon}=1\times10^{-2}\,s^{-1}$, (b) $\dot{\varepsilon}=3\times10^{-2}\,s^{-1}$, and (c) $\dot{\varepsilon}=1\times10^{-1}\,s^{-1}$ (tensile axis is horizontal) [140].

increased initial strain rates at a given strain. The increase in both the number and size of cavities with increased initial strain rates resulted in an increase in the cavity volume fraction. Quantitative measurement of cavitation indicated that the density, average size, and volume fraction of cavities tended to increase with increase in the initial strain rate from 1×10^{-2} to $1 \times 10^{-1}\ s^{-1}$ (Figures 7.6–7.8). This indicates that low strain rate is beneficial to retarding the nucleation and growth of cavities.

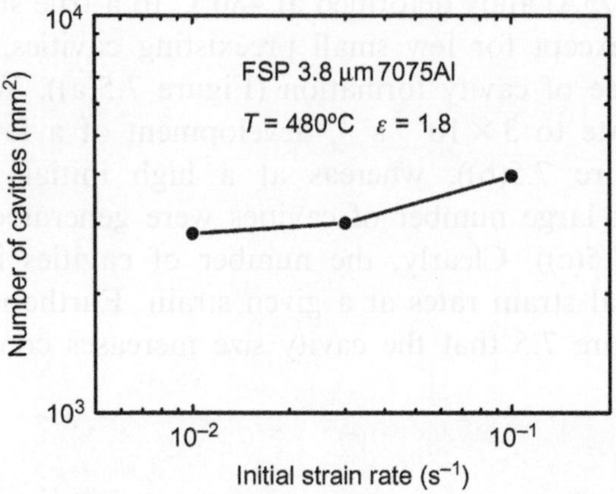

Figure 7.6 Variation of cavity density with initial strain rate for FSP 3.8 μm 7075Al alloy deformed at 480° C to a true strain of 1.8 [140].

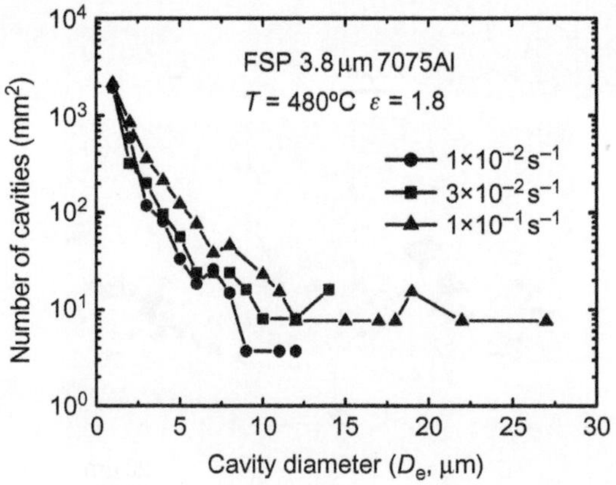

Figure 7.7 Variation of cavity density with cavity diameter for FSP 3.8 μm 7075Al alloy deformed at 480° C and different initial strain rates to a true strain of 1.8 [140].

The decrease in cavitation with strain rate is believed to result from a lower flow stress at a lower strain rate because flow stress influences the stresses built up at grain boundary and thus the tendency for cavity nucleation. A lower flow stress at lower strain rate means smaller stresses accumulated at grain boundaries. Moreover, at a lower strain rate, the stresses generated by GBS would have more time to be relaxed. Thus, superplastic deformation performed at a lower strain rate could reduce the local stresses at grain boundary irregularities, thereby decreasing the possibility of cavity nucleation.

7.2.2 Test Temperature

Figure 7.9 shows the effect of testing temperature on the cavitation of the FSP 3.8 μm 7075Al alloy deformed at an initial strain rate of $1 \times 10^{-2}\,s^{-1}$ to a true strain of 1.8. The cavity density tended to decrease with increasing temperature. However, the specimen deformed at 510°C exhibited the largest cavity size (Figure 7.9(c)). It is interesting to note that the specimen deformed at the optimum superplasticity temperature of 480°C exhibited the lowest volume fraction of cavities and ratio of large size cavities.

Quantitative measurement of cavitation indicated that the cavity density decreased continuously with increasing temperature from 450°C to 510°C (Figure 7.10). However, the specimen deformed at the optimum superplasticity temperature of 480°C exhibited the lowest

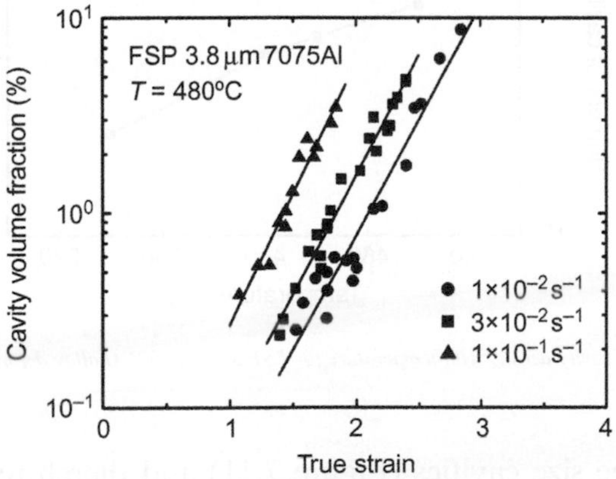

Figure 7.8 Variation of cavity volume fraction with true strain for FSP 3.8 μm 7075Al alloy deformed at 480° C and different initial strain rates [140].

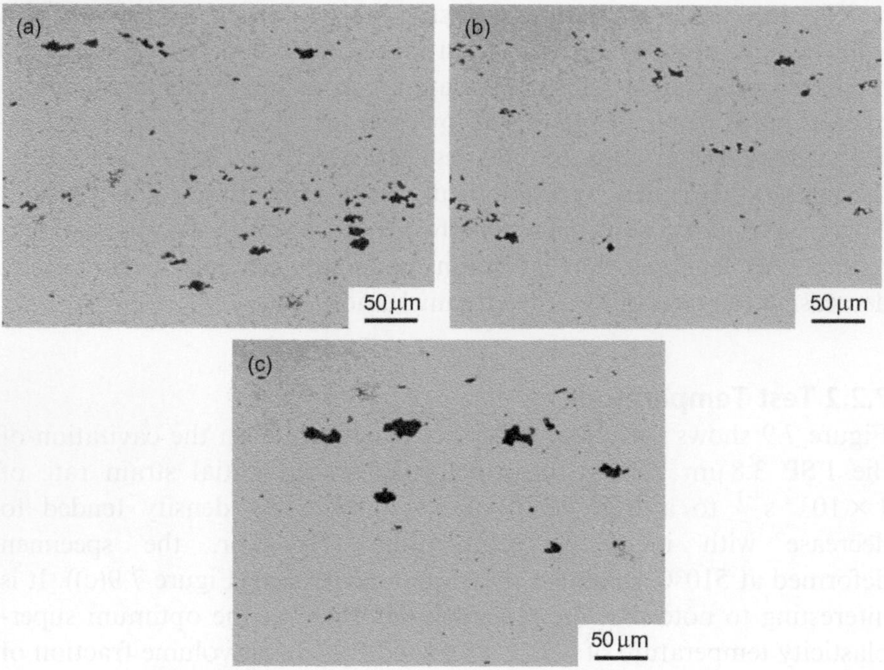

Figure 7.9 Optical micrographs showing cavitation of FSP 3.5 μm 7075Al alloy deformed at an initial strain rate of 1 × 10^{-2} s^{-1} to a true strain of 1.8: (a) 450°C, (b) 480°C, and (c) 510°C (tensile axis is horizontal) [140].

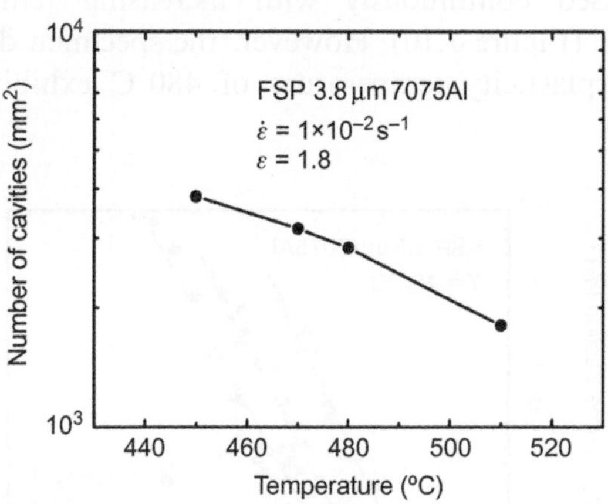

Figure 7.10 Variation of cavity density with temperature for FSP 3.8 μm 7075Al alloy deformed at 480°C to a true strain of 1.8 [140].

fraction of large size cavities (Figure 7.11) and therefore the least cavity fraction volume (Figure 7.12). Although the specimen deformed at 510°C exhibited the lowest cavity density, it had the highest fraction of large size cavities and cavity fraction volume.

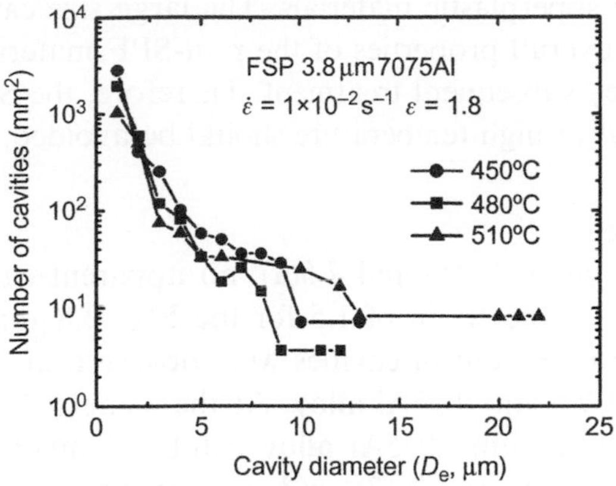

Figure 7.11 Variation of cavity density with cavity diameter for FSP 3.8 μm 7075Al alloy deformed at an initial strain rate of 1×10^{-2} s^{-1} and different temperatures to a true strain of 1.8 [140].

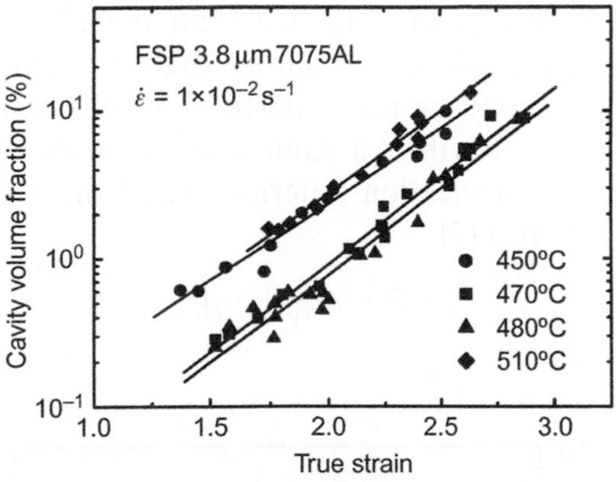

Figure 7.12 Variation of cavity volume fraction with true strain for FSP 3.8 μm 7075Al alloys deformed at an initial strain rate of 1×10^{-2} s^{-1} and different temperatures [140].

Increasing temperature generally reduces flow stress of a superplastic material, thereby reducing stress accumulation at grain boundaries. Moreover, at a higher temperature, stresses generated by GBS can be more rapidly relaxed by diffusion and/or dislocation processes. Therefore, it would be reasonable to expect that increasing deformation temperature would reduce cavity nucleation. The highest ratio of large size cavities and higher cavity volume fraction at 510°C for the FSP 7075Al are attributed to more intense grain growth at higher temperature exposure (Figure 3.8). As will be discussed in Section 7.2.3, the increase in the grain size results in an increase in cavitation tendency in the superplastic materials. The large size cavities are most harmful to the overall properties of the post-SPF materials and healed with difficulty by subsequent treatment. Therefore, the SPF of the FSP aluminum alloys at high temperature should be avoided.

7.2.3 Grain Size

As shown in Figures 7.1(a) and 7.5(a), no apparent cavity formation was revealed at a true strain of 1.5 for the FSP 3.8 μm 7075Al alloy, whereas a certain amount of cavities were observed at a true strain of 1.0 for the FSP 7.5 μm 7075Al alloy. At the same strain, for example $\varepsilon = 1.8$, the FSP 3.8 μm 7075Al alloy exhibited much lower density and size of cavities (Figure 7.13), and consequently much lower cavity volume fraction (Figure 7.7). This can be accounted for by the reduction in both grain and secondary phase particle sizes.

TEM examinations have indicated that the FSP 3.8 μm 7075Al alloy exhibited smaller precipitate size than the FSP 7.5 μm 7075Al alloy (Figure 2.3). Equation (7.1) shows that the refinement in both matrix grains and secondary phase particles results in minimizing the formation of cavities. Further, a similar trend is also predicted by a generally used cavity nucleation criterion based on stress equilibrium at cavity interface [116,117]

$$r_c = 2(\gamma + \gamma_p - \gamma_i)/\sigma \tag{7.2}$$

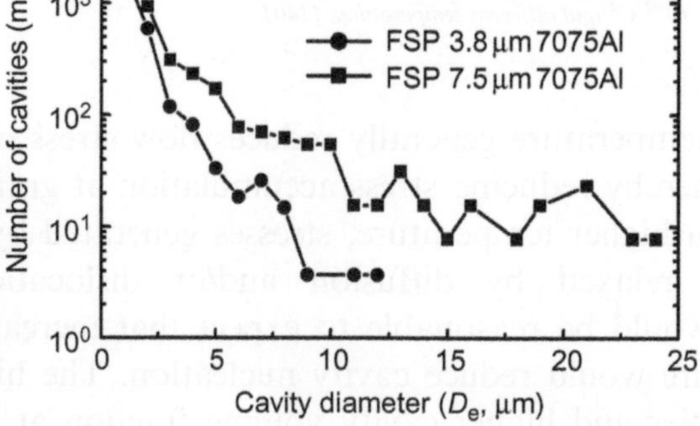

Figure 7.13 Variation of cavity density with cavity diameter for FSP 7.5 and 3.8 μm 7075Al alloys deformed at 480° C and an initial strain rate of 1×10^{-2} s^{-1} to a true strain of 1.8 [140].

where r_c is the critical void radius, γ is the interfacial energy per unit area of void surface, γ_p is the interfacial energy per unit area of particle surface, γ_i is the energy per unit area of particle/matrix interface, and σ is the applied stress. γ_p and γ_i are zero when no particle is involved in the cavity nucleation process.

Equation (7.2) predicts that the decrease in the applied stress results in an increase in the critical void radius, i.e., the cavity nucleation becomes difficult under low stress. Furthermore, since voids formed at particles by decohesion immediately assume the size of the particles, Eq. (7.2) suggests that stable cavity nucleation at smaller particles become impossible as the applied stress decreases.

Figure 3.6 shows that the flow stress of the FSP 3.8 μm 7075Al alloy is much lower than that of the FSP 7.5 μm 7075Al alloy. The lower flow stress associated with smaller grain and secondary phase particle sizes increase the critical size of a cavity, thereby making the nucleation event more difficult. Thus, the FSP 3.8 μm 7075Al alloy can endure more strain before cavity initiation than the FSP 7.5 μm 7075Al alloy.

7.3 CAVITY GROWTH MECHANISM AND CRITICAL STRAIN

It is well known that the stable nuclei or preexisting cavity grows during superplastic tensile flow through stress-directed vacancy diffusion and plastic deformation of the surrounding materials. With the increase of cavity radius, the contribution from diffusion is quickly reduced and the plasticity-controlled growth is dominant [115,118]. As shown in Figures 7.8, 7.12, and 7.14, the cavity volume fraction exhibited an exponential dependence on the true strain on log-linear scales for the two FSP 7075Al alloys under the investigated initial strain rates of 1×10^{-2}–$1\times10^{-1}\ s^{-1}$ and temperatures of 450–510°C. This implies that the cavity growth in the FSP 7075Al alloys is controlled by plasticity as suggested by Hancock [119]. Hence, the cavity volume fraction C_V can be expressed as

$$C_V = C_0 \exp(\eta\varepsilon) \quad (7.3)$$

where η is the cavity growth rate parameter, ε is the true strain, and C_0 is the preexponential constant. The values of η and C_0 determined

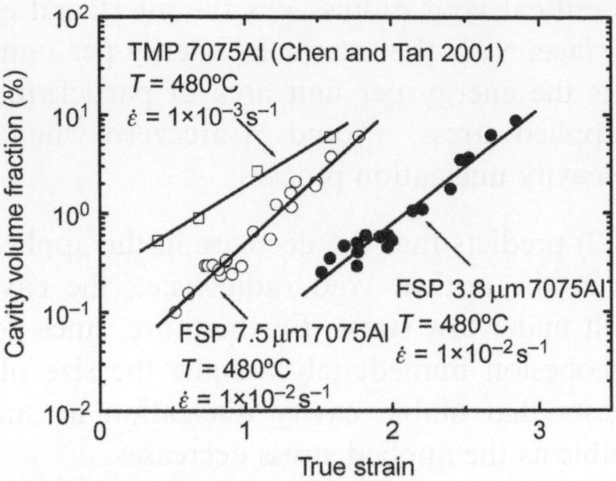

Figure 7.14 Variation of cavity volume fraction with true strain for FSP 7.5 and 3.8 μm 7075Al alloys deformed at 480°C and an initial strain rate of 1×10^{-2} s^{-1} [140].

Table 7.1 Cavity Growth Rate Parameter η, Preexponential Constant C_0, Geometric Factor k_s, and Nucleation Strain ε_0 in FSP 7075Al Alloys Superplastically Deformed at Various Temperatures and Strain Rates [140]

Materials	T (°C)	Strain Rate (s^{-1})	η	C_0	ε_0
3.8 μm 7075Al	450	1×10^{-2}	2.36	2.09×10^{-2}	
3.8 μm 7075Al	470	1×10^{-2}	2.76	3.38×10^{-3}	
3.8 μm 7075Al	480	1×10^{-1}	3.11	1.19×10^{-2}	0.94
		3×10^{-2}	2.82	5.59×10^{-3}	1.29
		1×10^{-2}	2.68	3.36×10^{-3}	1.53
3.8 μm 7075Al	510	1×10^{-2}	2.55	1.66×10^{-2}	
7.5 μm 7075Al	480	1×10^{-2}	3.15	2.11×10^{-2}	

by the linear regression are summarized in Table 7.1. Clearly, the cavity growth rate parameter η increases with increasing initial strain rate. This shows that the FSP 7075Al alloy exhibited faster cavity growth at higher strain rates.

Although Eq. (7.3) is widely accepted and used by numerous investigators [118,120–128], the physical meaning of C_0 in Eq. (7.3) is not well understood so far. It has been considered as a fitting preexponential constant [118,122,126] or the initial (or preexisting) cavity volume fraction in the superplastic materials before deformation [123,125,127,128]. However, there are two problems with Eq. (7.3) and C_0 when C_0 is associated with initial cavity level. First, different C_0 values were obtained by linear regression from experimental data for the FSP 3.8 μm 7075Al specimens deformed at different initial strain rates of $1 \times 10^{-2} - 1 \times 10^{-1}\ s^{-1}$ (Table 7.1). If C_0 is the initial cavity level, it would imply that three different values of initial cavity volume fraction are obtained in the FSP 3.8 μm 7075Al alloy. For a superplastic material deformed at same temperature, it is impossible for it to have different initial cavity levels. Second, Eq. (6.6) predicts that when true strain >0, cavitation will develop. However, when a superplastic material is deformed, it is more reasonable to assume that cavitation does not occur below a critical strain. For the FSP 3.8 μm 7075Al alloy deformed at 480°C and an initial strain rate of $1 \times 10^{-2}\ s^{-1}$, OM examinations showed that no apparent cavitation occurred at a true strain of <1.5 (Figure 7.5(a)). Similarly, TEM examinations by Jiang et al. [3] revealed that there was no cavity nucleation in a TMP 7075Al alloy deformed at 510°C and $8.3 \times 10^{-4}\ s^{-1}$ to a strain level of 300%. Clearly, the C_0 determined by linear regression from experimental data cannot be simply considered the initial cavity level. A more appropriate form of cavitation growth equation should be used to exactly describe the cavitation behavior of superplastic materials.

The basic relationship for plasticity-controlled cavity growth is [129]

$$dV/d\varepsilon = KV \tag{7.4}$$

where V is the volume of the cavity and K is a parameter. Assuming that K is a constant for a given material and testing condition, it results in

$$C_V = C_{Vo} \exp[\eta(\varepsilon - \varepsilon_0)] \tag{7.5}$$

where the cavity growth rate parameter η is identical to K and C_{Vo} is the cavity volume fraction at strain ε_0. Equation (7.3) is a simplified form of Eq. (7.5) with an assumption that $\varepsilon_0 = 0$.

In previous studies, several investigators adopted Eq. (7.5) to describe the cavitation behavior [130–133]. While Taleff et al. [131] considered ε_0 as a reference strain, Nieh et al. [132] and Caceres and Wilkinson [133] took ε_0 as the nucleation strain for stable cavity formation. For the latter case, C_{Vo} is the cavity volume fraction at the nucleation strain ε_0. However, it seems that the nucleation strain cannot exactly describe the real situation about cavitation initiation due to two reasons. First, there are usually preexisting cavities in superplastic materials as in the present FSP 7075Al alloy. Above a certain strain, it is likely for these preexisting cavities to grow. Second, because of limit of resolution of optical microscope, some small nucleated cavities are not visible. At a large strain, these small cavities will grow. In this study, ε_0 is defined as critical strain below which no cavitation occurs.

For the FSP 7075Al alloys, clearly, there exists a critical strain for stable cavity formation as evidenced by Figure 7.5(a). Therefore, it is appropriate to adopt Eq. (7.5) to describe the cavitation behavior of superplastic FSP 7075Al alloys. For this purpose, the cavity level in the grip region of deformed specimens was experimentally measured by means of OM, and this value is referred to as the initial cavity volume fraction, i.e., the cavity volume fraction at the critical strain ε_0. The grip regions from three FSP 3.8 μm 7075Al specimens deformed at 480°C and different initial strain rates of $1 \times 10^{-2} - 1 \times 10^{-1}$ s^{-1} were examined, and more than 35 micrographs were taken throughout each grip region. The initial cavity volume fraction is determined to be ~0.2%. The initial cavities are believed to result from rolling, FSP, as well as polishing.

In Figure 7.15, the cavity data of the FSP 3.8 μm 7075Al alloy specimens deformed at 480°C and different initial strain rates of

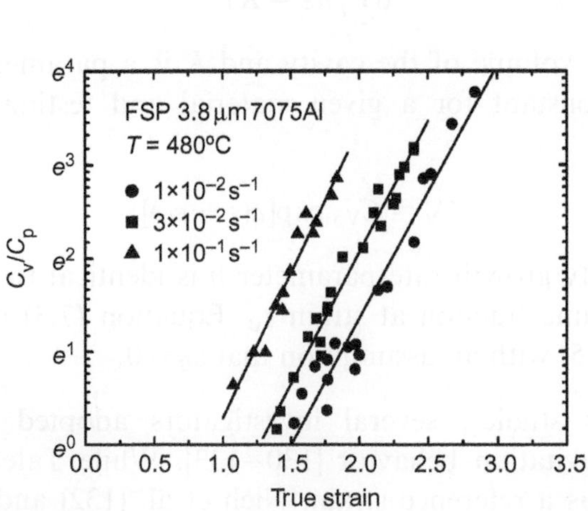

Figure 7.15 Variation of normalized cavity volume fraction with true strain for FSP 3.8 μm 7075Al alloy deformed at 480°C and different initial strain rates [140].

$1 \times 10^{-2} - 1 \times 10^{-1}$ s^{-1} is replotted as cavity volume fraction normalized by the experimentally determined initial cavity volume fraction, C_{V_0}, versus true strain. Extrapolating linear regression lines to e^0, i.e., $C_V/C_{V_0} = 1$, generates the values of the nucleation strain under different initial strain rates, and these values are summarized in Table 7.1. For the FSP 3.8 μm 7075Al alloy specimens deformed at 480°C and an initial strain rate of 1×10^{-2} s^{-1}, a nucleation strain of 1.53 is obtained. This value is quantitatively consistent with OM examinations (Figure 7.5(a)) where no apparent cavitation was revealed at a true strain of 1.5.

A true strain of 1.5 corresponds to an elongation of 348%. This means that no cavitation occurs in the FSP 3.8 μm 7075Al alloy deformed at 480°C and an initial strain rate of 1×10^{-2} s^{-1} to an elongation of 300%. Considering that most forming operations require a ductility of <200%, the present FSP 7075Al alloys can provide excellent post-SPF properties. Furthermore, Table 7.1 shows that the critical strain increases with decreasing initial strain rate. A linear relationship is observed between the critical strain and initial strain rates for the FSP 3.8 μm 7075Al alloy deformed at 480°C (Figure 7.16)

$$\varepsilon_0 = 0.35 - 0.60 \log \dot{\varepsilon} \tag{7.6}$$

where ε_0 is the critical strain below which no cavitation occurs and $\dot{\varepsilon}$ is the initial strain rate.

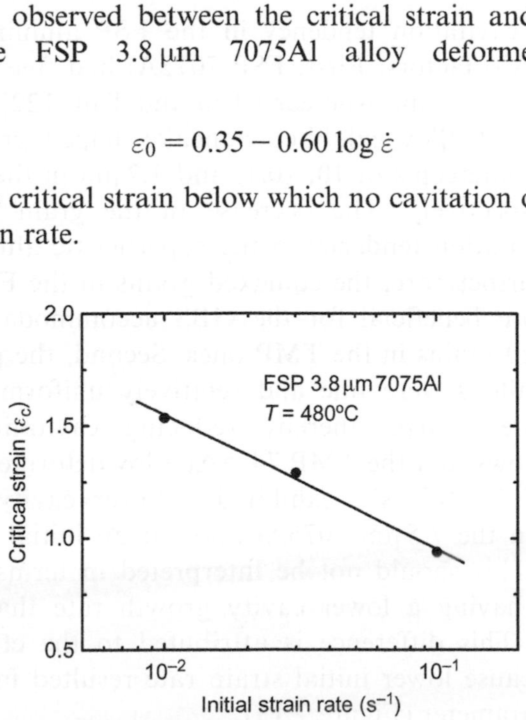

Figure 7.16 Variation of nucleation strain, ε_0, with initial strain rate for FSP 3.8 μm 7075Al alloy deformed at 480°C [140].

7.4 COMPARISON BETWEEN CAVITATION BEHAVIORS OF FSP AND TMP ALUMINUM ALLOYS

FSP and TMP aluminum alloys exhibited different microstructural features. Therefore, it is worthwhile to compare the superplastic cavitation behavior of two types of aluminum alloys to provide an insight to the advantages of FSP over TMP. Unfortunately, no cavitation data of TMP 7xxx series aluminum alloys is available at high initial strain rates of $\geq 1 \times 10^{-2}\ s^{-1}$. For comparison, cavitation data of Chen and Tan [122] for TMP 7475Al alloy at 480°C and a low initial strain rate of $1 \times 10^{-3}\ s^{-1}$ were plotted in Figure 7.14. Clearly, the cavity volume fraction of the two FSP 7075Al alloys is much lower than that of the TMP 7475Al alloy at a given strain, even though the FSP alloys were deformed at a higher initial strain rate of $1 \times 10^{-2}\ s^{-1}$. Considering the fact that increasing initial strain rate resulted in an increased cavity fraction volume as shown in Figure 7.8 and discussed in Section 7.2.3, the FSP aluminum alloy exhibits a significantly reduced cavitation level compared to the TMP alloy.

The lower cavitation tendency in the FSP aluminum alloys is attributed to two factors. First, FSP 7075Al had fine and equiaxed grains of 3.8 and 7.5 μm, whereas Chen and Tan [122] reported that the TMP 7475Al alloy exhibited pancake-shaped grains with the average linear intercepts of 10, 10.1, and 7.2 μm in the L, T, and N directions, respectively. The decrease in the grain size tends to reduce the cavitation tendency in the superplastic aluminum alloys (Eq. (7.1)). Furthermore, the equiaxed grains in the FSP aluminum alloys are more beneficial for the GBS accommodation than the pancake-shaped grains in the TMP ones. Second, the precipitates in FSP 7075Al alloys were fine and relatively uniformly distributed throughout the matrix, thereby reducing cavitation tendency. Figure 7.14 shows that the TMP 7475Al alloy deformed at an initial strain rate of $1 \times 10^{-3}\ s^{-1}$ exhibited a lower cavity growth rate parameter than the 7.5 μm 7075Al alloy at an initial strain rate of $1 \times 10^{-2}\ s^{-1}$. This should not be interpreted in terms of the TMP 7475Al alloy having a lower cavity growth rate than the 7.5 μm 7075Al alloy. This difference is attributed to the effect of initial strain rate because lower initial strain rate resulted in lower cavity growth rate parameter (Figure 7.15).

The study on the nucleation strain for superplastic cavitation initiation is very limited in the literature. Jiang et al. [3] examined the cavitation of TMP 7075Al alloy by means of TEM. It was revealed that there was no cavity nucleation in TMP 7075Al alloy deformed at 510°C and an initial strain rate of $8.3 \times 10^{-4}\ s^{-1}$ to a strain level of 300%. The nucleation strain for cavitation initiation in the FSP 3.8 μm 7075Al alloy was determined to be ~1.5 by both OM examinations (Figure 7.5(a)) and normalization by the initial cavity volume fraction (Table 7.1). A true strain of 1.5 corresponds to an elongation of 348%. This shows that both FSP and TMP aluminum alloys exhibit a similar nucleation strain for cavitation initiation. However, it should be pointed out that the FSP alloy was deformed at a higher initial strain rate of $1 \times 10^{-2}\ s^{-1}$. Considering the significant effect of the strain rate on the nucleation strain (Figure 7.13), it can be concluded that the FSP aluminum alloy exhibits a higher value of nucleation strain than TMP one.

CHAPTER 8

Superplastic Forming of Friction Stir Processed Plates

Conventional superplasticity is based on starting sheet material of <3 mm thickness. This limitation mainly comes from the fact that the processing of superplastic sheets requires thermomechanical processing in the form of controlled hot rolling to obtain fine-grained microstructure with grain size <15 μm. So applications of friction stir processing can be visualized at several different levels:

1. Enhancing the properties locally to achieve high strain rate superplastic forming and/or low temperature superplastic forming
2. Use friction stir processing to expand superplastic forming of thicker sheets and plates
3. Use friction stir processing for superplastic forming of low-cost cast plates
4. Use of friction stir processing for expanding to superplastic punch forming and superplastic forging.

Much of this book has presented data on a variety of aluminum alloys to show the advantages of FSP in grain refinement where it results in high strain rate superplasticity and/or low temperature superplasticity. Mishra and Mahoney [134–136] have originally detailed the concepts and then reviewed the work initially. Figure 8.1 shows a gas formed cone of 7475Al alloy formed in 18 min at ~1 MPa (150 psi). The conventionally processed sheet took 95 min to fully form at the same gas pressure. This shows the extent of enhancement in kinetics of superplastic flow. In this example the sheet was processed by 14 overlapping passes to refine the microstructure in the entire target region for forming. An interesting idea that builds from the ability of FSP to refine microstructure locally is "selective superplastic forming." Figure 8.2 shows a superplastically formed component from thin sheet. Note that only certain regions of the sheet have undergone superplastic deformation to form the features in the component. For such components, conceptually the region of interest can be selectively processed to refine the grain size and subsequent gas forming will form that

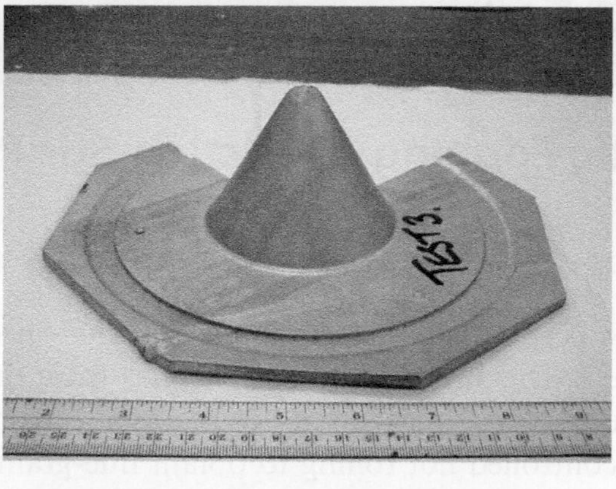

Figure 8.1 Example of gas pressure cone test for FSP 7475Al ~1 MPa (150 psi), 18 min.

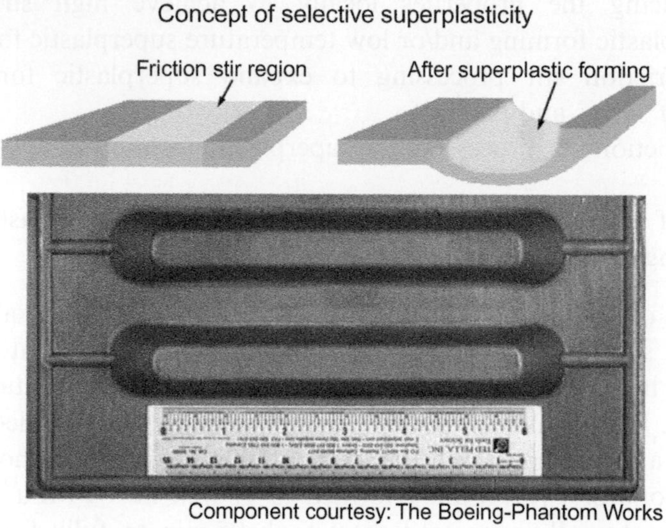

Figure 8.2 A schematic of the concept of selective superplasticity for a component along with an example component that has only a fraction of sheet undergoing superplastic deformation. Note that since only a part of the overall sheet goes through superplastic deformation and that region can be friction stir processed for forming.

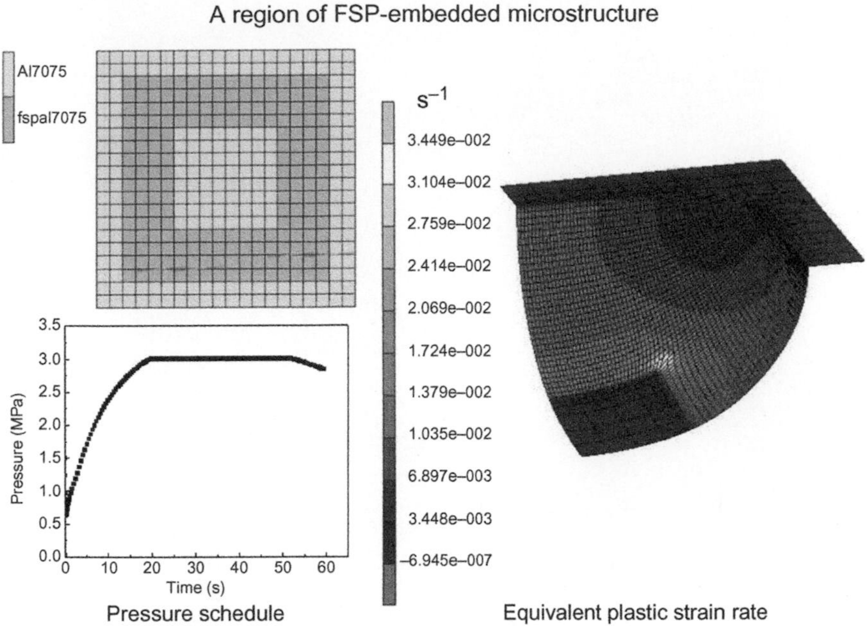

Figure 8.3 Finite element modeling of selective superplasticity using data for FSP 7075Al. The strain rates in the friction stir processed region are in high strain rate superplasticity domain.

region. Figure 8.3 shows a finite element simulation of selective superplastic forming using 7075Al data that was presented in Chapter 3. Note that the strain rate during forming does get into high strain rate superplastic domain and the gas pressure was applied for just 1 min in this simulation. The maximum stress was equivalent to 3 MPa.

CHAPTER 9

Potential of Extending Superplasticity to Thick Sections

A unique feature of friction stir processing is that the grain refinement is achieved without changing the geometrical shape of the sheet or plate. Designers can take advantage of this aspect to design components out of thicker sheets and plates which have not been considered in the past. This does require rethinking of manufacturing approach of certain components. Figure 9.1 illustrates superplastic tensile elongation of ~800% at 460°C following FSP in 5-mm-thick FSP 7075 Al. The thickness limit for superplasticity has not been established but results up to 12-mm-thick specimens have been published. Figure 9.2 shows one such example from Mahoney et al. reported in Ref. [136] for 7475Al alloy. In this case the strain rates were slow, but the full extent of superplastic domain in such thick specimens has not been established. In some cases, the limit of gas pressure can become translated into the limiting forming strain rate because of the flow stress.

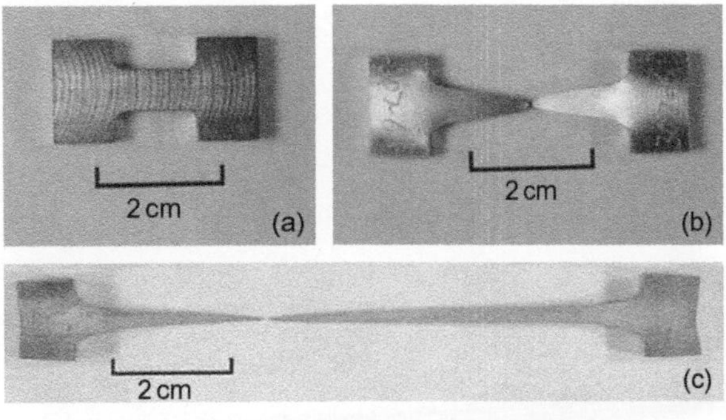

Figure 9.1 (a) 5-mm-thick tensile sample, (b) limited tensile elongation and severe necking without FSP, and (c) 800% superplastic elongation in 5-mm-thick FSP 7075 Al. Taken from Ref. [136].

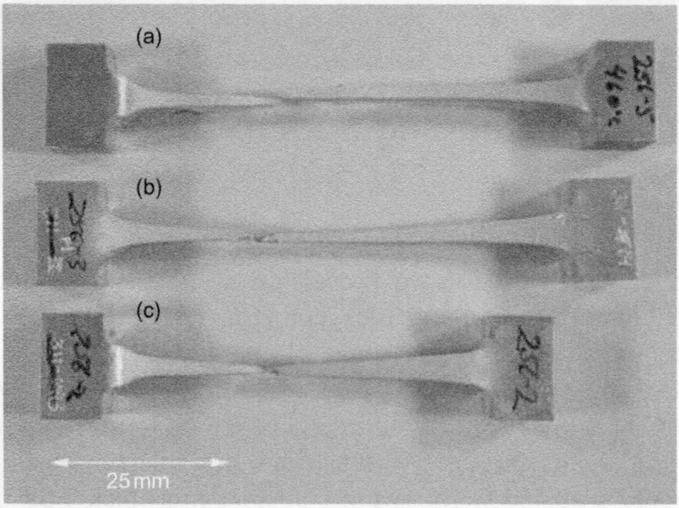

Figure 9.2 Superplastic strain in 12-mm-thick friction stir processed 7475 Al at $2 \times 10^{-4}\ s^{-1}$: (a) 460° C, 670% strain, (b) 440° C, 630% strain, and (c) 420° C, 470% strain. Taken from Ref. [136].

CHAPTER 10

Potential of Superplastic Forming of Low-Cost Cast Plate

One of the most cost-effective ways of producing sheets is twin-roll casting. In twin-roll casting, some level of high temperature deformation can be imparted while the material moved through the solidification rolls. For the concept of selective superplastic forming, this can become a very effective way of producing low cost starting material. Alternatively it is also possible to think of direct-chill cast plates as the starting material. In Chapters 3–5, examples were included of traditional cast alloy like A356, and cast Al–Zn–Mg–Sc alloy which would normally be further thermomechanically processed to obtain wrought microstructure. In both these cases, friction stir processing resulted in high strain rate superplasticity. Johannes et al. [137] reported on enhanced superplasticity in continuous cast 5083Al alloy sheet. These show excellent possibility of taking a low-cost starting material and convert selected regions into superplastic regions for subsequent gas forming.

CHAPTER 11

Superplastic Punch Forming and Superplastic Forging

Superplastic punch forming and closed die forging are other possibilities because of the ability to convert thick sheets and plates into superplastic microstructure. Dutta et al. [138] reported on punch forming of 9-pass FSP 7075Al with a thickness of 5 mm. They used the approach of limiting dome height (LDH) to establish formability. Punch forming tests were carried out at 400°C and 450°C. While punch forming tests at 400°C were performed at only one strain rate, i.e., at $10^{-2}\ s^{-1}$, three different strain rates were employed at 450°C, as higher elongations and larger cup depths could be obtained at this temperature. The maximum cup depth of 52 mm, as shown in Figure 11.1, was achieved at a slow strain rate of $10^{-3}\ s^{-1}$ at a temperature of 450°C consistent with the high tensile elongation under these conditions. They also conducted stage-wise measurements of thickness strain and compared with finite element modeling (FEM) simulation at 450°C and $10^{-2}\ s^{-1}$. The progress of deformation and the change of shape of the blank at three stages of forming under the above parameters have been demonstrated in Figure 11.2 through FEM mesh distortions. The cross-sectioned photographs of two stages have been shown in Figure 11.2(b) and (c), while Figure 11.3(a) shows a lower LDH of only 18.25 mm for the as-received 7075Al alloy for comparison.

During the punch forming period, higher biaxial tensile stresses in the crown initiate formation of cavities. At the pole the additional compressive stress in the thickness direction (σ_3) prevented initiation/growth of cavities, as shown in Figure 11.4. Cavities were observed at points away from the pole towards equator and especially near the outer surface due to greater stretching forces. All the samples deformed to the final failure stage had cracked at a distance away from pole. However, when the specimen fractures before this stage as in the case of as-received 7075Al alloy, the pole was observed to be the thinnest. The micrographs at different regions of the cup formed at 723 K, $10^{-2}\ s^{-1}$, have been shown in Figure 11.4. The identical grain sizes of

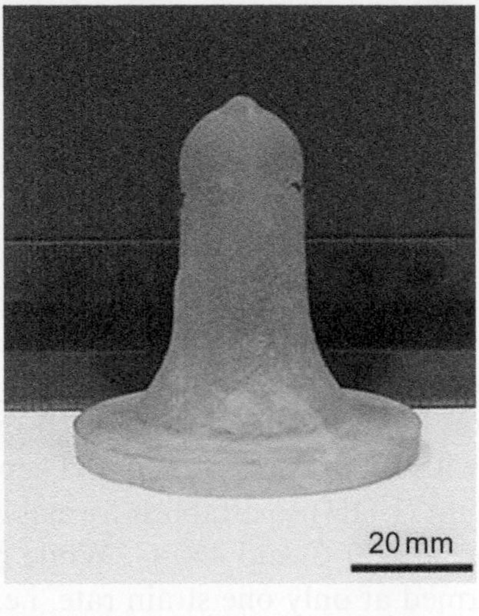

Figure 11.1 Large cup depth of 52 mm at failure after punch forming of FSP 7075Al alloy at 450° C, with a strain rate of 10^{-3} s^{-1}.

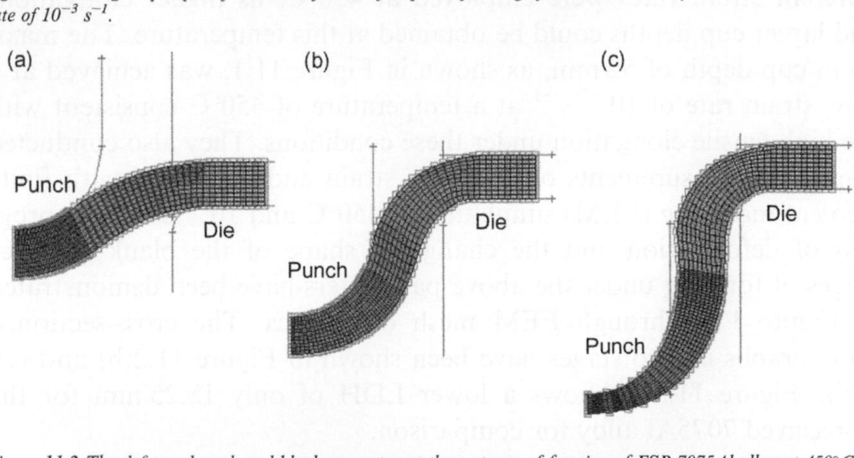

Figure 11.2 The deformed mesh and blank geometry at three stages of forming of FSP 7075Al alloy at 450° C at punch strokes of (a) 8.17, (b) 18.5, and (c) 28.5 mm.

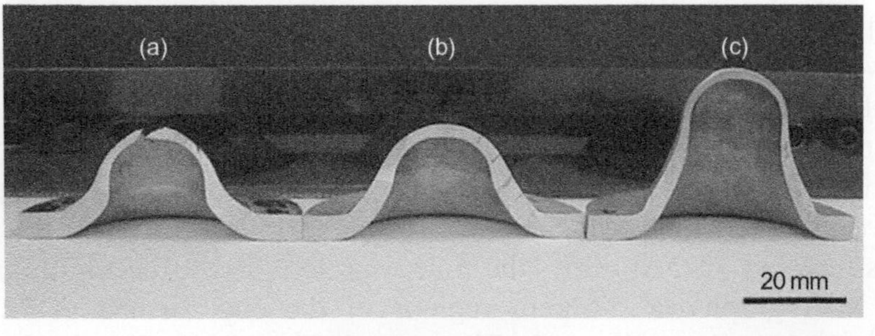

Figure 11.3 Three cups punch formed at 450° C: (a) as-received 7075Al alloy, (b) FSP 7075Al alloy at 18.5 mm punch stroke, and (c) FSP 7075Al alloy at 28.5 mm punch stroke.

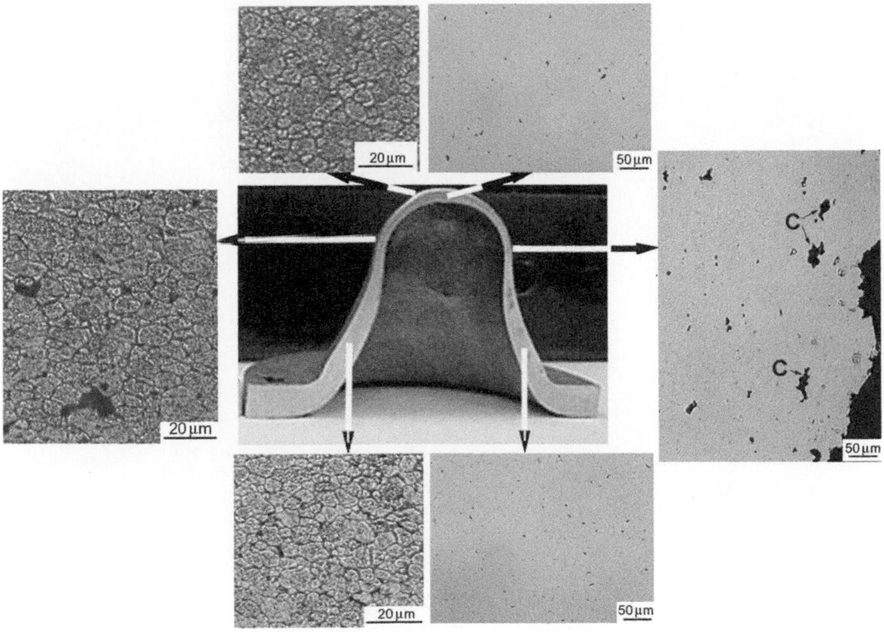

Figure 11.4 Concentration of cavities, mostly at the outer surface of the necked region after formation of a cup to the depth of 28.5 mm at 723 K, at a strain rate of 10^{-2} s^{-1}. Microstructures at different regions indicate negligible grain coarsening.

the flange and the initial FSP sheet indicate absence of static grain growth during the short period of test (~180 s).

Similar performance in closed die forging can be expected because of the ease of material flow in superplastic conditions. Many friction stir welding work has been reported for 25-mm-thick plates. Such plates can be processed with a raster pattern to create stock material for forging with superplastic properties.

CHAPTER 12

Friction Stir Welding and Superplastic Forming for Multisheet Structures

One of the key benefits of superplastic forming (SPF) is the ability to form multisheet components. These integrally stiffened structures or sandwich structures are used in many aerospace applications. Most of the initial examples were developed for titanium alloys as they can be diffusion bonded easily. The simple approach of diffusion bonding/superplastic forming (DB/SPF) involved DB of sheets with masked areas and subsequent gas forming by applying pressure in between the sheets. Very complex structures could be obtained by using three, four, or five sheets. Aluminum alloys on the other hand have surface oxide film that hindered easy DB. This can be overcome with friction stir welding (FSW) of sheets in lap joint configuration with patterns. FSW retains the fine-grained microstructure of superplastic sheet. The work of Grant et al. [139] has demonstrated the feasibility of making multisheet structures by combining FSW and friction stir spot welding (FSSW) with SPF. Figure 12.1 shows an example of a three-sheet structure created by FSW through two and three sheets. Fusion welding of aluminum alloys leads to a complete loss of superplasticity in the welded region because of microstructural changes. As noted earlier, the FSW microstructure consists of fine grains and superplastic properties are not degraded. A new opportunity involves microstructural tailoring by controlled heat input during FSW. The grain size can be varied by changing the thermal input. By controlling the microstructure, one can make the superplastic flow stress of the FSW region lower, higher, or equal to the parent sheet. Grant et al. [139] have also used FSSW to create different types of multisheet structures. Their work is opening up new possibilities of sandwich structures using aluminum alloy sheets.

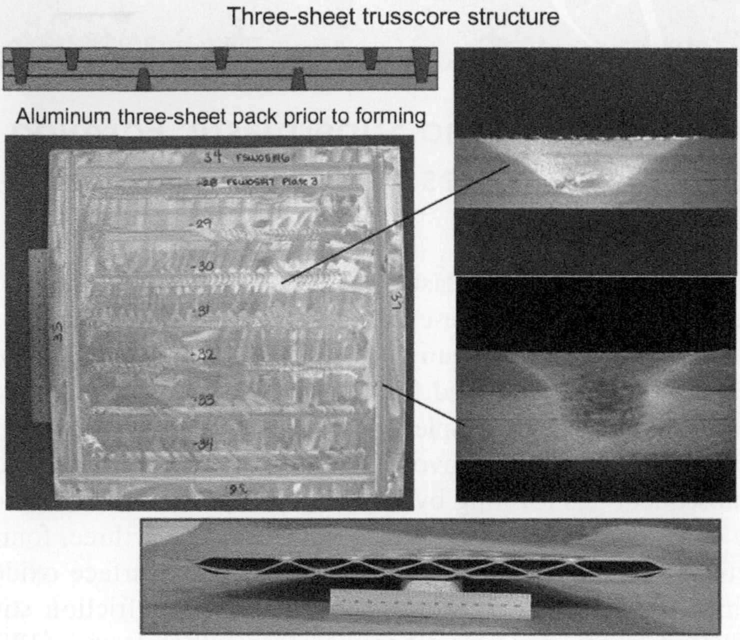

Figure 12.1 An example of multisheet structure created by a combination of FSW and superplastic forming. Figure courtesy: Glenn Grant.

CHAPTER 13

Summary

Unitized structures are very attractive for metallic structures as they reduce component count and reduce cost. High strain rate superplasticity is critical to reduction of forming time during superplasticity while low temperature superplasticity is very promising for reduction of energy consumption during manufacturing. Friction stir processing (FSP) results in a unique microstructure with fine grain size and large fraction of high-angle grain boundaries. These microstructural features lead to high strain rate superplasticity in many commercial aluminum alloys. Even cast microstructure can be converted to superplastic microstructure in a single pass. Large regions can be covered with rastering patterns. The critical strain for onset of cavitation is higher for friction stir processed microstructure. This is related to break up and homogenization of constituent particles, which typically cause nucleation of cavities, during FSP. The thickness of sheet or plate does not change during FSP. This attribute of FSP can be used to extend superplastic forming to thick sections, and to design components with the idea of selective superplasticity using low-cost cast or wrought plates. Conversion of thick sheets and plates as a stock material with superplastic properties can be utilized for superplastic punch forming and superplastic forging. These new possibilities expand the design space for superplastically formed components.

REFERENCES

[1] N.E. Paton, C.H. Hamilton, J. Wert, M. Mahoney, JOM 34 (1982) 21–27.

[2] X.G. Jiang, Q.L. Wu, J.Z. Cui, L.X. Ma, Metall. Trans. A 24 (1993) 2596–2598.

[3] X.G. Jiang, J.Z. Cui, L.X. Ma, Acta Metall. Mater. 41 (1993) 2721–2727.

[4] K. Higashi, M. Mabuchi, in: Superplasticity in Advanced Materials, ICSAM'97, vol. 243-2, 1997, pp. 267–276.

[5] H.P. Pu, F.C. Liu, J.C. Huang, Metall. Mater. Trans. A 26 (1995) 1153–1166.

[6] R.S. Mishra, T.R. Bieler, A.K. Mukherjee, Acta Metall. Mater. 43 (1995) 877–891.

[7] S. Komura, P.B. Berbon, M. Furukawa, Z. Horita, M. Nemoto, T.G. Langdon, Scr. Mater. 38 (1998) 1851–1856.

[8] R.Z. Valiev, R.K. Islamgaliev, I.V. Alexandrov, Prog. Mater. Sci. 45 (2000) 103–189.

[9] I.C. Hsiao, J.C. Huang, Scr. Mater. 40 (1999) 697–703.

[10] T.R. Mcnelley, R. Crooks, P.N. Kalu, S.A. Rogers, Mater. Sci. Eng., A 166 (1993) 135–143.

[11] I.C. Hsiao, J.C. Huang, Metall. Mater. Trans. A 33 (2002) 1373–1384.

[12] S. Komura, Z. Horita, M. Furukawa, M. Nemoto, T.G. Langdon, Metall. Mater. Trans. A 32 (2001) 707–716.

[13] K.T. Park, D.Y. Hwang, S.Y. Chang, D.H. Shin, Metall. Mater. Trans. A 33 (2002) 2859–2867.

[14] R.Z. Valiev, D.A. Salimonenko, N.K. Tsenev, P.B. Berbon, T.G. Langdon, Scr. Mater. 37 (1997) 1945–1950.

[15] S. Ota, H. Akamatsu, K. Neishi, M. Furukawa, Z. Horita, T.G. Langdon, Mater. Trans. 43 (2002) 2364–2369.

[16] P.B. Berbon, N.K. Tsenev, R.Z. Valiev, M. Furukawa, Z. Horita, M. Nemoto, et al., Metall. Mater. Trans. A 29 (1998) 2237–2243.

[17] R.S. Mishra, R.Z. Valiev, S.X. McFadden, R.K. Islamgaliev, A.K. Mukherjee, Philos. Mag. A 81 (2001) 37–48.

[18] M. Noda, M. Hirohashi, K. Funami, Mater. Trans. 44 (2003) 2288–2297.

[19] N. Tsuji, K. Shiotsuki, Y. Saito, Mater. Trans. JIM 40 (1999) 765–771.

[20] Z.Y. Ma, Metall. Mater. Trans. A 39A (2008) 642–658.

[21] R.S. Mishra, Z.Y. Ma, Mater. Sci. Eng., R 50 (2005) 1–78.

[22] R.S. Mishra, M.W. Mahoney, S.X. McFadden, N.A. Mara, A.K. Mukherjee, Scr. Mater. 42 (1999) 163–168.

[23] R.S. Mishra, M.W. Mahoney, in: Superplasticity in Advanced Materials, ICSAM'00, vol. 357-3, 2001, pp. 507–512.

[24] Z.Y. Ma, R.S. Mishra, M.W. Mahoney, Acta Mater. 50 (2002) 4419–4430.

[25] Z.Y. Ma, R.S. Mishra, M.W. Mahoney, R. Grimes, Mater. Sci. Eng., A 351 (2003) 148–153.

[26] Z.Y. Ma, R.S. Mishra, M.W. Mahoney, R. Grimes, Metall. Mater. Trans. A 36A (2005) 1447–1458.

[27] Z.Y. Ma, R.S. Mishra, Scr. Mater. 53 (2005) 75–80.

[28] Z.Y. Ma, R.S. Mishra, M.W. Mahoney, Scr. Mater. 50 (2004) 931–935.

[29] I. Charit, R.S. Mishra, Mater. Sci. Eng., A 359 (2003) 290–296.

[30] I. Charit, R.S. Mishra, J. Mater. Res. 19 (2004) 3329–3342.

[31] I. Charit, R.S. Mishra, Acta Mater. 53 (2005) 4211–4223.

[32] M.W. Mahoney, C.G. Rhodes, J.G. Flintoff, R.A. Spurling, W.H. Bingel, Metall. Mater. Trans. A 29 (1998) 1955–1964.

[33] M. Mahoney, R.S. Mishra, T. Nelson, J. Flintoff, R. Islamgaliev, Y. Hovansky, in: Friction Stir Welding and Processing, 2001, pp. 183–194.

[34] Glossary of terms used in metallic superplastic materials, JISH 7007, Japanese Standards Association, Tokyo, 1995.

[35] H. Huang, Q. Wu, J. Hua, Superplasticity and Superplastic Forming, TMS, Warrendale, PA, 1988.

[36] Y. Liu, G. Yang, J. Lian, X. Zeng, Superplasticity in Metals, Ceramics and Intermetallics, Materials Research Society, Pittsburgh, PA, 1990.

[37] W. Zheng, Z. Baoliang, J. Mater. Sci. Lett. 13 (1994) 1806–1808.

[38] K. Matsuki, T. Aida, J. Kusui, in: Towards Innovation in Superplasticity II, vol. 304-3, 1999, pp. 255–260.

[39] K. Matsuki, H. Sugahara, T. Aida, N. Takatsuji, J. Kusui, K. Yokoe, Mater. Trans. JIM 40 (1999) 737–743.

[40] S. Lee, T.G. Langdon, Mater. Res. Soc. Symp. Proc. 601 (2000) 359–364.

[41] Z. Horita, S. Lee, S. Ota, K. Neishi, T.G. Langdon, in: Superplasticity in Advanced Materials, ICSAM'00, vol. 357-3, 2001, pp. 471–476.

[42] J.W. Edington, K.N. Melton, C.P. Cutler, Prog. Mater. Sci. 21 (1976) 63–170.

[43] O.N. Senkov, M.M. Myshlyaev, Acta Metall. Mater. 34 (1986) 97–106.

[44] M.K. Rabinovich, V.G. Trifonov, Acta Mater. 44 (1996) 2073–2078.

[45] F. Li, D.H. Bae, A.K. Ghosh, Acta Mater. 45 (1997) 3887–3895.

[46] D.H. Shin, Y.J. Joo, C.S. Lee, K.T. Park, Scr. Mater. 41 (1999) 269–274.

[47] R. Kaibyshev, E. Avtokratova, A. Apollonov, R. Davies, Scr. Mater. 54 (2006) 2119–2124.

[48] P.S. Bate, F.J. Humphreys, N. Ridley, B. Zhang, Acta Mater. 53 (2005) 3059–3069.

[49] J. Liu, D.J. Chakrabarti, Acta Mater. 44 (1996) 4647–4661.

[50] J.K. Mackenzie, Biometrika 45 (1958) 229–240.

[51] F.C. Liu, P. Xue, Z.Y. Ma, Mater. Sci. Eng., A 547 (2012) 55–63.

[52] J.Q. Su, T.W. Nelson, C.J. Sterling, J. Mater. Res. 18 (2003) 1757–1760.

[53] H.J. Frost, M.F. Ashby, Deformation Mechanism Maps, Pergamon Press, Oxford, 1982.

[54] F.C. Liu, Z.Y. Ma, L.Q. Chen, Scr. Mater. 60 (2009) 968–971.

[55] F.C. Liu, Z.Y. Ma, Scr. Mater. 58 (2008) 667–670.

[56] F. Musin, R. Kaibyshev, Y. Motohashi, G. Itoh, Scr. Mater. 50 (2004) 511–516.

[57] C. Xu, M. Furukawa, Z. Horita, T.G. Langdon, Acta Mater. 51 (2003) 6139–6149.

[58] D.L.e. Zalensas, Aluminum Casting Technology, second ed., 1993. p. 77

[59] D.L. Zhang, L. Zheng, Metall. Mater. Trans. A 27 (1996) 3983–3991.

[60] T. Din, J. Campbell, Mater. Sci. Technol. Ser. 12 (1996) 644–650.

[61] Y.B. Yu, P.Y. Song, S.S. Kim, J.H. Lee, Scr. Mater. 41 (1999) 767–771.

[62] T.G. Nieh, J. Wadsworth, T. Imai, Scr. Metall. Mater. 26 (1992) 703–708.

[63] M. Mabuchi, K. Higashi, Philos. Mag. Lett. 70 (1994) 1–6.

[64] B.Q. Han, K.C. Chan, T.M. Yue, W.S. Lau, Scr. Metall. Mater. 33 (1995) 925–930.

[65] J.C. Kim, Y. Nishida, H. Arima, T. Ando, Mater. Lett. 57 (2003) 1689–1695.

[66] S. Shivkumar, S. Ricci, C. Keller, D. Apelian, J. Heat Treat. (1990) 63.

[67] A.K. Mukherjee, Grain Boundaries in Engineering Materials, Claiton Publishing, Baton Rouge, LA, 1975.

[68] A.K. Mukherje, J.E. Bird, J.E. Dorn, ASM Trans. Q. 62 (1969) 155.

[69] T.R. Bieler, A.K. Mukherjee, Mater. Sci. Eng., A 128 (1990) 171–182.

[70] W.J. Kim, K. Higashi, J.K. Kim, Mater. Sci. Eng., A 260 (1999) 170–177.

[71] T.G. Nieh, J. Wadsworth, in: Superplasticity in Advanced Materials, ICSAM'97, vol. 243-2, 1997, pp. 257–266.

[72] H. Iwasaki, T. Mori, M. Mabuchi, K. Higashi, Metall. Mater. Trans. A 29 (1998) 677–683.

[73] G.Q. Tong, K.C. Chan, Mater. Sci. Eng., A 286 (2000) 218–224.

[74] W.J. Kim, O.D. Sherby, Acta Mater. 48 (2000) 1763–1774.

[75] O.D. Sherby, P.M. Burke, Prog. Mater. Sci. 13 (1967) 323–390.

[76] K.L. Murty, F.A. Mohamed, J.E. Dorn, Acta Metall. Mater. 20 (1972) 1009–1018.

[77] H. Oikawa, K. Sugawara, S. Karashima, Trans. Jpn Inst. Metals 19 (1978) 611–616.

[78] T.G. Nieh, R. Kaibyshev, L.M. Hsiung, N. Nguyen, J. Wadsworth, Scr. Mater. 36 (1997) 1011–1016.

[79] A. Arieli, A.K. Mukherjee, Mater. Sci. Eng. 45 (1980) 61–70.

[80] A. Ball, M.W. Hutchinson, Metal Sci. J. 3 (1969) 1–7.

[81] J. Weertman, J.R. Weertman, Constitutive Relations and their Physical Basis, Riso National Laboratory, Denmark, 1987.

[82] P. Yavari, T.G. Langdon, Acta Metall. Mater. 30 (1982) 2181–2196.

[83] E.M. Taleff, P.J. Nevland, S.J. Yoon, Deformation, Processing, and Properties of Structural Materials, 2000, pp. 373–384.

[84] H. Fukuyo, H.C. Tsai, T. Oyama, O.D. Sherby, ISIJ Int. 31 (1991) 76–85.

[85] T.G. Nieh, L.M. Hsiung, J. Wadsworth, R. Kaibyshev, Acta Mater. 46 (1998) 2789–2800.

[86] O.D. Sherby, J. Wadsworth, Prog. Mater. Sci. 33 (1989) 169–221.

[87] S. Primdahl, A. Tholen, T.G. Langdon, Acta Metall. Mater. 43 (1995) 1211–1218.

[88] A.H. Chokshi, Mater. Sci. Eng., A 166 (1993) 119–133.

[89] T.G. Langdon, in: N. Ridley (Ed.), Superplasticity: 60 Years after Pearson, Institute of Materials, London, 1994, p. 9.

[90] W.A. Rachinger, J. Inst. Metals 81 (1952) 33–41.

[91] T.G. Langdon, Acta Metall. Mater. 42 (1994) 2437–2443.

[92] R.B. Vastava, T.G. Langdon, Acta Metall. Mater. 27 (1979) 251–257.

[93] T.G. Langdon, J. Mater. Sci. 16 (1981) 2613–2616.

[94] P. Shariat, R.B. Vastava, T.G. Langdon, Acta Metall. Mater. 30 (1982) 285–296.

[95] T.G. Langdon, Metall. Trans. 3 (1972) 797.

[96] T.G. Langdon, J. Mater. Sci. 41 (2006) 597–609.

[97] T. Watanabe, Mater. Sci. Eng., A 166 (1993) 11–28.

[98] M. Kawasaki, T.G. Langdon, Mater. Trans. 49 (2008) 84–89.

[99] H. Kokawa, T. Watanabe, S. Karashima, Philos. Mag. A 44 (1981) 1239–1254.

[100] F.A. Mohamed, J. Mater. Sci. Lett. 7 (1988) 215–217.

[101] P.K. Chaudhury, F.A. Mohamed, Acta Metall. Mater. 36 (1988) 1099–1110.

[102] A.F. Norman, I. Brough, P.B. Prangnell, Aluminium Alloys: Their Physical and Mechanical Properties, Pts 1–3, vol. 331-3, 2000, pp. 1713–1718.

[103] T.R. McNelley, M.E. McMahon, Metall. Mater. Trans. A 28 (1997) 1879–1887.

[104] M. Eddahbi, T.R. McNelley, O.A. Ruano, Metall. Mater. Trans. A 32 (2001) 1093–1102.

[105] L.B. Johannes, R.S. Mishra, Mater. Sci. Eng., A 464 (2007) 255–260.

[106] T.R. Mcnelley, D.J. Michel, A. Salama, Scr. Metall. 23 (1989) 1657–1662.

[107] R.D. Schellen, G.H. Reynolds, Metall. Trans. 4 (1973) 2199–2203.

[108] N. Ridley, D.W. Livesey, A.K. Mukherjee, Metall. Trans. A 15 (1984) 1443–1450.

[109] P. Bompard, J.Y. Lacroix, A. Varloteaux, Aluminum 64 (1988) 162.

[110] M.J. Stowell, D.W. Livesey, N. Ridley, Acta Metall. Mater. 32 (1984) 35–42.

[111] V.M. Segal, Mater. Sci. Eng., A 197 (1995) 157–164.

[112] R.Z. Valiev, A.V. Korznikov, R.R. Mulyukov, Mater. Sci. Eng., A 168 (1993) 141–148.

[113] D.H. Shin, K.T. Park, Mater. Sci. Eng., A 268 (1999) 55–62.

[114] A.H. Chokshi, Towards Innovation in Superplasticity I, vol. 233-2, 1997, pp. 89–108.

[115] D.H. Bae, A.K. Ghosh, Acta Mater. 50 (2002) 511–523.

[116] A.K. Ghosh, in: H. Hanson, A. Horsewell, T. Leffers, E.H. Lilholt (Eds.), Deformation of Polycrystals: Mechanisms and Microstructures, Riso National Laboratory, Roskilde, Denmark, 1982, p. 277.

[117] A.H. Chokshi, T.G. Langdon, Acta Metall. Mater. 35 (1987) 1089–1101.

[118] N. Ridley, Z.C. Wang, Towards Innovation in Superplasticity I, vol. 233-2, 1997, pp. 63–80.

[119] J.W. Hancock, Metal. Sci. 10 (1976) 319.

[120] J. Pilling, N. Ridley, Acta Metall. Mater. 34 (1986) 669–679.

[121] H. Iwasaki, T. Mori, K. Higashi, Towards Innovation in Superplasticity II, vol. 304-3, 1999, pp. 675–680.

[122] C.L. Chen, M.J. Tan, Mater. Sci. Eng., A 298 (2001) 235–244.

[123] J.J. Blandin, B. Hong, A. Varloteaux, M. Suery, G. LEsperance, Acta Mater. 44 (1996) 2317–2326.

[124] K.C. Chan, K.K. Chow, Mater. Lett. 56 (2002) 38–42.

[125] A.K. Ghosh, D.H. Bae, in: Superplasticity in Advanced Materials, ICSAM'97, vol. 243-2, 1997, pp. 89–98.

[126] H. Iwasaki, M. Mabuchi, K. Higashi, Acta Mater. 49 (2001) 2269–2275.

[127] H.Y. Wu, Mater. Sci. Eng., A 291 (2000) 1–8.

[128] P.D. Nicolaou, S.L. Semiatin, A.K. Ghosh, Metall. Mater. Trans. A 31 (2000) 1425–1434.

[129] M.J. Stowell, Met. Sci. 14 (1980) 267–272.

[130] J. Lian, M. Suery, Mater. Sci. Technol. Ser. 2 (1986) 1093–1098.

[131] E.M. Taleff, T. Leon-Salamanca, R.A. Ketcham, R. Reyes, W.D. Carlson, J. Mater. Res. 15 (2000) 76–84.

[132] T.G. Nieh, D.R. Lesuer, C.K. Syn, Mater. Sci. Eng., A 202 (1995) 43–51.

[133] C.H. Caceres, D.S. Wilkinson, Acta Metall. Mater. 32 (1984) 423–434.

[134] R.S. Mishra, Friction stir processing for superplasticity, Adv. Mater. Processes 162(2) (2004) 45–47.

[135] R.S. Mishra, M.W. Mahoney, Metal superplasticity enhancement and forming process. US Patent 6,712,916, March 30, 2004.

[136] R.S. Mishra, M.W. Mahoney, ISBN-13: 978-0-87170-840-3. in: R.S. Mishra, M.W. Mahoney (Eds.), Friction Stir Welding and Processing, ASM International, 2007, p. 309.

[137] L.B. Johannes, I. Charit, R.S. Mishra, R. Verma, Mater. Sci. Eng., A 464 (2007) 351–357.

[138] A. Dutta, I. Charit, L.B. Johannes, R.S. Mishra, Mater. Sci. Eng., A 395 (2005) 173–179.

[139] G.J. Grant, D. Herling, W. Arbegast, C. Allen, C. Degen, in: 2006 International Conference on Superplasticity in Advanced Materials, Chengdu, China, June 23, 2006.

[140] Z.Y. Ma, R.S. Mishra, Acta Mater. 51 (2003) 3551–3569.

[141] F.C. Liu, Z.Y. Ma, Scripta Mater. 62 (2010) 125–128.

[142] Z.Y. Ma, F.C. Liu, R.S. Mishra, Acta Mater. 58 (2010) 4693–4704.

[143] I. Charit, R.S. Mishra, M.W. Mahoney, Scripta Mater. 47 (2002) 631–636.